Raw Veganism

Human beings are getting fatter and sicker. As we question what we eat and why we eat it, this book argues that living well involves consuming a raw vegan diet.

With eating healthfully and eating ethically being simpler said than done, this book argues that the best solution to health, environmental, and ethical problems concerning animals is raw veganism—the human diet. The human diet is what humans are naturally designed to eat, and that is, a raw vegan diet of fruit, tender leafy greens, and occasionally nuts and seeds. While veganism raises challenging questions over the ethics of consuming animal products, while also considering the environmental impact of the agriculture industry, raw veganism goes a step further and argues that consuming cooked food is also detrimental to our health and the environment. Cooking foods allows us to eat food that is not otherwise fit for human consumption and in an age that promotes eating foods in 'moderation' and having 'balanced' diets, this raises the question of why we are eating foods that should only be consumed in moderation at all, as moderation clearly implies they aren't good for us. In addition, from an environmental perspective, the use of stoves, ovens and microwaves for cooking contributes significantly to energy consumption and cooking in general generates excessive waste of food and resources. Thus, this book maintains that living well and living a noble life, that is, good physical and moral health, requires consuming a raw vegan diet.

Exploring the scientific and philosophical aspects of raw veganism, this novel book is essential reading for all interested in promoting ethical, healthful, and sustainable diets.

Carlo Alvaro is Adjunct Assistant Professor of Philosophy at New York City College of Technology, USA. He is the author of *Ethical Veganism, Virtue Ethics, and the Great Soul* (2019).

Routledge Studies in Food, Society and the Environment

Localizing Global Food
Short Food Supply Chains as Responses to Agri-Food System Challenges
Edited by Sophia Skordili and Agni Kalfagianni

Seafood Supply Chains
Governance, power and regulation
Miriam Greenwood

Civil Society and Social Movements in Food System Governance
Edited by Peter Andrée, Jill K. Clark, Charles Z. Levkoe and Kristen Lowitt

Voice and Participation in Global Food Politics
Alana Mann

Plant-Based Diets for Succulence and Sustainability
Edited by Kathleen May Kevany

Sustainable Food System Assessment
Lessons from Global Practice
Edited by Alison Blay-Palmer, Damien Conaré, Ken Meter, Amanda Di Battista and Carla Johnston

Raw Veganism
The Philosophy of the Human Diet
Carlo Alvaro

The Bioeconomy Approach
Constraints and Opportunities for Sustainable Development
Udaya Sekhar Nagothu

For more information about this series, please visit: http://www.routledge.com/books/series/RSFSE/

Raw Veganism
The Philosophy of the Human Diet

Carlo Alvaro

First published 2020
by Routledge
2 Park Square, Milton Park, Abingdon, Oxon OX14 4RN

and by Routledge
52 Vanderbilt Avenue, New York, NY 10017

Routledge is an imprint of the Taylor & Francis Group, an informa business

© 2020 Carlo Alvaro

The right of Carlo Alvaro to be identified as author of this work has been asserted by him in accordance with sections 77 and 78 of the Copyright, Designs and Patents Act 1988.

All rights reserved. No part of this book may be reprinted or reproduced or utilised in any form or by any electronic, mechanical, or other means, now known or hereafter invented, including photocopying and recording, or in any information storage or retrieval system, without permission in writing from the publishers.

Trademark notice: Product or corporate names may be trademarks or registered trademarks, and are used only for identification and explanation without intent to infringe.

British Library Cataloguing-in-Publication Data
A catalogue record for this book is available from the British Library

Library of Congress Cataloging-in-Publication Data
Names: Alvaro, Carlo, 1974- author.
Title: Raw veganism : the philosophy of the human diet / Carlo Alvaro.
Description: Abingdon, Oxon ; New York, NY : Routledge, 2021. |
Series: Routledge studies in food, society and the environment |
Includes bibliographical references and index. |
Identifiers: LCCN 2019048492 | ISBN 9780367435028 (hardback) |
ISBN 9780367435394 (paperback) | ISBN 9781003003960 (ebook)
Subjects: LCSH: Veganism--Philosophy. | Raw foods. | Nutrition. |
Diet--Moral and ethical aspects. | Food of animal origin--Moral
and ethical aspects.
Classification: LCC HV4711 .A293 2021 | DDC 179/.3–dc23
LC record available at https://lccn.loc.gov/2019048492

ISBN: 978-0-367-43502-8 (hbk)
ISBN: 978-0-367-43539-4 (pbk)
ISBN: 978-1-003-00396-0 (ebk)

Typeset in Bembo
by Taylor & Francis Books

 Printed in the United Kingdom by Henry Ling Limited

For Malaika

For Maisie

Contents

Preface viii
Acknowledgements x

Introduction 1
1 The ethics of veganism 6
2 There is more than animal suffering 28
3 In vitro meat 43
4 Ethical veganism: What it is, what it is not, and what it should be 60
5 Raw veganism: The human diet 69
6 Education and abolition 89
7 Raw veganism and children 107
8 Conclusion 126

References 130
Index 145

Preface

When I wrote the initial draft of my first book, *Ethical Veganism, Virtue Ethics, and the Great Soul*, I expressly intended it for an academic audience. Proud of my scholarly work, I sent the manuscript to my publisher—and they told me that it was not academic enough. Between my initial draft and the process of academization, it took me two years to complete the book. I did enjoy the process, though those two years were possibly the most intense of my life. While I was writing the book, I was also writing several articles, teaching full-time, and I was (and still am) a full-time dad.

So, at the end of the spring semester of 2018, after completing my book, I decided to take a break. A philosopher's break, mind you, does not mean not writing at all; it just means writing articles and reading a million books. One of the articles I was working on is about the significance of animal suffering or, to be more precise, the idea that all philosophers writing in defense of animals end up grounding their respective arguments on the notion of animal suffering. That is, eating animals is wrong. Why? Because they suffer. Using animals for scientific research is wrong. Why? Because they suffer. Go vegan! Why? Because—you guessed it—animals suffer. I do believe that animals suffer. But animal suffering, believe it or not, is not the principal reason why I do not eat animals. What are these reasons, then? Well, I don't want to spoil it for you. I explain these reasons in "There is more than animal suffering," which is Chapter 2 of this book.

Also, I was writing another article, which deals with evolution of our species and human diet. This paper turned out to be longer than a normal paper, and it became "Raw veganism: The human diet", the central chapter of this book. Again, I do not want to spoil it for you. This paper was the impetus to the ideas that culminated in the creation of this book, *Raw Veganism: The Philosophy of the Human Diet*.

At this point, I was still enjoying my philosopher's break, though I started considering writing a new book on raw veganism. This time, I did not want to write a book for academics. I wanted to speak to a wider audience and tell them about my experience as a raw vegan, and the moral, environmental, and dietary benefits of raw veganism. I wanted to write a popular book. So I did (or at least I thought I did). I sent the proposal to my publisher—and they told me that it was too academic!

Well, *ecce libro*. This time I wrote a scholarly book that also speaks to a larger audience. *Raw Veganism* contains provocative discussions about the abolition of animal products, veganarchism, animal ethics, food ethics, environmental ethics, nutrition, and evolutionary biology that may be of interest to physicians, nutritionists, policy makers, and philosophers. *Raw Veganism* shows that, contrary to what most people believe, cooking food hinders, rather than promotes, good health. Also contrary to popular belief, our ancestors did not feast upon animal flesh. And no, the fact that we have canine teeth means nothing. Humans are fruit eaters—that's the human diet. Moreover, eating fruit is what made our brains grow larger and made us humans. Cooking food, especially animal-derived, is one of the worst incidents that happened to humanity.

All animal species that exist, or have existed, have a specific diet, a diet on which a species is adapted to thrive. Well, you can see where this is going. Where is the evidence? There is tons of evidence. The problem is that the notion of eating a diet of fruit and salad is so wacky and foreign to humans today that scientists have overlooked the evidence or have not connected the dots or have deliberately avoided connecting the dots, or all of the above. There is too much interest in maintaining the status quo. Although I consider myself an optimist, I don't have high hopes for a raw vegan world. But my job as a philosopher is the search for truth, regardless of how wacky of uncomfortable the truth may be. And the truth is that cooking food and eating animals are mistakes. Our species-specific food is fruit.

We can run from the truth, but we can't hide—cooked food, especially animal-based, grains, soda, oil, alcoholic beverages, coffee, sugar, and salt are not human food. The human diet is fresh, uncooked, unprocessed fruit, with the addition of tender, leafy greens and a handful of nuts and seeds. Cooked food is dead. Cooking food denatures food. The human body requires water-rich and live food. The human diet provides all the nutrients that the human body requires. Why would it not? Our ancestors thrived on such a diet for millions of years prior to agriculture, cooking, and hunting.

The human diet makes the question of sustainability redundant. It avoids violence toward animals. It improves the digestion. It makes obesity unheard of. It confers mental clarity and physical strength. It extends life. It avoids the excessive use of detergents and disinfectants. It avoids the use of natural gas as a fuel to cook. It avoids zoonoses, infectious diseases caused by bacteria and parasites that spread between animals and humans. It can bring us back to nature and make us appreciate its beauty and importance. It makes us see the wrongness of killing animals and polluting the environment. My hope, therefore, is that this book will inspire people and make them seriously consider raw veganism.

Acknowledgements

First, I would like to thank my editor, Hannah Ferguson, for embracing the vision of this book, and for choosing its beautiful cover. Also, the staff at Routledge, particularly John Baddeley, have been wonderful. I owe a very special thanks to Jan Deckers. Thank you for your valuable work in environmental ethics, and thank you for taking the time to read through the manuscript and offering thoughtful criticisms and suggestions for improvement. I am exceedingly grateful to Robert C. Jones for taking an interest in my ideas, for giving me helpful feedback, and for being supportive of my career. Thanks, Cheryl Abbate and Laura Wright, for your contributions to animal ethics and to veganism. Thank you, Gregory Tague, for giving me a platform to speak and for sharing my ideas and for being so supportive. I thank Rocco Alvaro for our discussions about Stoicism, life, and raw veganism. Finally, my greatest thanks, as always, are to my family. I am thankful to my wife Malaika for her loving support over the years, and for reading and discussing my project and giving me helpful comments and ideas. And I am thankful to my sons George and Jon and my daughter Valentina. You are a constant inspiration, and the living proof that being vegans since birth is not hard, and it is a healthful and moral way to live.

Introduction

Recent reports on world population health are both alarming and disheartening. To put it candidly and succinctly, human beings are getting fatter and sicker. Considering my beloved country, for example, The National Health and Nutrition Examination Survey (NHANES) reports that obesity rates among US adults in 2015–2016 was a frightening 39.8%.[1] Since 1999, when the figure was 30.5%, obesity rates have been progressively increasing—and it does not look like they are going to get any lower any time soon. A recent study estimates that there will be "65 million more obese adults in the USA and 11 million more obese adults in the UK by 2030."[2] Equally alarming (though not surprising), according to the American Heart Association's Heart and Stroke Statistics nearly half of US adults suffer from some form of cardiovascular disease.[3] An interesting fact is that while humans are getting sicker and sicker, wild animals do not have any obesity or cardiovascular problems.

According to the Centers for Disease Control and Prevention (CDC), Americans are eating more calories than they did 40 years ago. This steady increase in calories, CDC suggests, is due to an increased availability of food, especially high-calorie foods. "The increase in calories is mainly due to an increase in carbohydrate consumption."[4] A worrisome fact is that most Americans do not eat enough fruit and vegetables;[5] and meat consumption is also steadily increasing in the USA (also in other countries).[6] I don't believe it is difficult to infer the conclusion to this argument, although many researchers and the public constantly deny it. But if I have to state it clearly, human beings eat the wrong food.[7] Due to many factors, from historical reasons to psychological ones, human beings have abandoned their natural way of living and eating. The term 'natural' is a dangerous and contentious word in philosophy. I don't want to be accused of the naturalistic fallacy here. I am not using the term to make a deep philosophical point. I simply want to suggest that, after all, living a simpler life, exercising more, and being in contact with nature is more beneficial to humans than the stress, processed food, and pollution of modern life. Why believe that humans should have a natural diet? Simply put, all animal species that exist and ever have existed on Earth have their respective natural diet. They consume as much as they want and do not get sick. In fact, only pets and liminal animals can suffer from obesity.[8] Again, the explanation is that those animals consume food that is not optimal to them. It is true that humans can survive on different diets. However, I believe that there are compelling reasons—beginning with the frightening reports just mentioned above—to show that cooked diets in general, especially meat-based diets, are not optimal human diets. What is the human diet? It is what humans are naturally designed to eat, and that is, a raw vegan diet of fruit, tender leafy greens, and occasionally nuts and seeds.

Consuming cooked food, especially animal-based, have rendered human beings sick. I predict resistance here. Some may argue, "No, no, no, it is not cooking food or eating animal products that make us sick and fat." What is it then? Why is the world population health declining? Surely, people are not eating enough fruits and vegetables, they eat cooked diets, including junk food (which is cooked food), and most people in the world eat meat and animal products. What causes humans to get sicker and sicker, then? Is it the fruit and vegetables? Obviously, it is not that. It is cooking food, because cooking food increases its calories and the variety of food that can be eaten.[9] Moreover, cooking enables humans to eat food that is not fit to human consumption. This, of course, has helped many humans survive through scarcity. But that is what cooked food essentially is, starvation food. One group of foods in particular that has taken a toll on human health is starches. As a group of researchers notes,

> Growing evidence shows that many of the chronic health conditions in developed countries could be prevented or moderated by dietary changes. The most common starchy foods in the United States diet, including white bread, cakes, and noodles, consist of a large percentage of highly digestible starch. There is concern that such rapidly digested starches may contribute to chronic disease in people and animals.[10]

Sure, the immediate objection is that it is the refined grains that are unhealthful, but whole grains are healthful. While it is generally true that some studies purport to show that consuming whole grains can be beneficial, the point is that all grains, including whole grains, are deficient in a number of important nutrients. The nutrition profile of whole grains (and cooked foods in general) is no match for fresh, uncooked fruit and greens, which "have the highest nutrient density score."[11] Also, in order to be digested, whole grains must be cooked, which further destroys any nutritional value. They are hard to digest and can damage the intestinal tract. Most importantly, all types of grains are virtually tasteless, and hence require the addition of sugar, salt, and fats.

Another immediate objection is that we do not need to resort to a raw diet to fight obesity and other chronic diseases. We can have a "balanced" diet or eat cooked food in moderation (oh, if only I had a nickel for every time I heard that!). But that begs the question. What is a balanced diet and what does "moderation" mean? With their ring of truth, these clichés have persuaded many people. Many things are harmful even in moderation. Should we use drugs in moderation? And regarding a balanced diet, do elephants, rabbits, or zebras eat a balanced diet and in moderation? It is evident that when animals eat their proper diet, the questions of moderation balance are moot. Such labels make nutrition more complicated than it actually is. I find the recommendation to eat in moderation quite suspicious. Why should anyone eat food that must be consumed in moderation? Most animals eat in abundance (when they can manage to find food). That is because they eat their specific food. If we are told that we have to consume certain foods in moderation, it follows that those foods are not entirely beneficial to our health. Cooked food is high in calories and low in nutrients. It seems quite obvious that processing food by cooking it cannot be as beneficial as consuming fresh, whole fruit, and tender leafy greens.

Furthermore, human diets generate moral issue involving our treatment of animals and nature. Anyone who is even remotely concerned about nature knows that animal agriculture is a leading cause of tremendous environmental degradation. Thus, many have suggested as a solution to global health and environmental problems a move in the direction of ethical veganism. While I agree that embracing ethical veganism is more ethical and more environmentally friendly than using animals for food, still it is not the most ethical and the most environmentally friendly approach. After all, cooking requires the use of ovens, stovetops, microwave ovens, which use coal, wood, gas, and electricity. Not to mention that production of food in factories also requires energy consumption. Compare that with raw vegan diets, which require no cooking, little to no dishwashing as well as the use of detergents. Furthermore, eating raw diets generates mostly organic, biodegradable waste that can be used as natural fertilizer. Thus, in this book I argue that the best solution to health, environmental, and ethical problems concerning animals is raw veganism—the human diet. I discuss the human diet from the point of view of nutrition science and evolution. I argue that living well and living a noble life, that is, good physical and moral health, require consuming a raw vegan diet.

In Chapter 1, I examine some moral approaches to ethical veganism. Historically, various forms of utilitarianism and deontological ethics have been used to frame the discussion and to defend veganism and vegetarianism. What I suggest in this chapter is that starting from a preferred moral theory (especially a theory that offers a certain procedure to determine what to do) and moving to the conclusion of veganism or vegetarianism is not the best option. I suggest that the most promising way to embrace veganism is to live a virtuous life. By virtuous life I mean cultivating moderation, courage, integrity, a sense of aesthetic, and the avoidance of violence. Using animals for food is antithetical to such virtues. An individual who possesses such virtues is one who lives simply and refuses violence, un-aesthetic values, and self-deception.

Animal rights advocates argue that eating animals is wrong because animals suffer. I agree that animals suffer and causing gratuitous suffering is wrong. But what if we discovered that animals do not suffer at all? What if Descartes was right in saying that animals behave like, but are not, sentient creatures? This is where my attitude toward nature differs from the attitude of many animal ethicists. I do believe that sentience is important, but it's not the principal factor. I would not eat or use animal products even if it were discovered that animals do not feel pain. Rather, I would still avoid such products because the actions and practices required to produce them are bloody, violent, unnecessary, intemperate, unaesthetic, and deleterious to our health. The foregoing are the themes of Chapter 2.

Modern life has alienated us from nature. In Chapter 3, I discuss the project of in vitro meat. In vitro meat, a.k.a. lab-grown meat, a.k.a. synthetic meat, and many other monikers, is believed by some to be the solution to our environmental problems. Here I discuss how the project of producing in vitro meat in fact stems from lack of moderation, and clear understanding of the role of food in human life. Furthermore, as I point out, even if in vitro meat researchers could overcome several difficulties involved in the production of synthetic meat, it would still be necessary to raise animals in order to produce animal byproducts. Ultimately, I point out that the project of in vitro meat is destined to be yet another option on the meat menu, and it will help more company profit than the environment.

Naturally, since the human diet is a diet devoid of animal products, another main theme of this book is ethical veganism. There are several books that discuss veganism as a dietary approach, and a handful others that discuss it as an ethical practice or philosophy. In Chapter 4, I discuss the notion of veganism: What it is, what it isn't, and what it should be. I argue that veganism should be an expression of virtue. The acquisition of certain basic virtue, such as nonviolence, moderation, and aesthetic sense lead to the embracement of the human diet, that is, a raw vegan diet, which is optimal for all humans.

In Chapter 5, I discuss some compelling evidence, including nutritional and evolutionary evidence, that a raw vegan diet is humans' ideal diet. Eating directly the food that nature produces is the best possible diet for human beings. What I mean by human diet is a diet that emphasizes fresh fruit, with the addition of tender leafy greens, some nuts and seeds, and excludes of tea, coffee, alcohol, salt, and sugar, and all cooked and processed food. Cooking food changes its molecular structure—and not for the better. It destroys nutrients, creates acrylamides and other carcinogenic substances, and denatures proteins, which can lead to many problems. Just to mention one problem: Leukocytosis is an increase in the body of white blood cells. This occurs as a reaction to inflammations or infections. In other words, when the body detects a threat, as a response it produces more white blood cells. This obviously does not happen when we eat fruit and salad. However, it does happen whenever we consume any type of cooked food. There is a wealth of scientific research showing that cooked food—vegan or not—shortens our lives.

Another interesting fact is to consider that humans have been around for about 150–200,000 years (not to mention that human-like creatures have been around for millions of years). During this time, no significant change occurred to our digestive system that equipped us for digesting cooked food. Thus, considering that human beings are evolved creatures, adapted to their environment, it is obvious that there is a diet that is specific and optimal for our species. Cooking food is a relatively new practice for humans. For the longest time, humans have eaten fruit and tender leafy greens. This is a scientifically documented fact—humans are frugivores. Consequently, cooking food is in no way beneficial to human health. The only benefit is that it provides easy calories by heat-processing food that otherwise would be indigestible.

In Chapter 6, I propose a valiant solution to the environmental issues in the way of moral education and environmental awareness and a qualified ban on animal-based products. In Chapter 7, I discuss the social and health implications of raising children on a raw vegan diet. Here I argue that veganism is not just nutritionally adequate, but rather nutritionally optimal for children. I especially discuss some of the health concerns regarding micronutrient deficiencies (B12, D, and more). Also, I show that the social lives of vegan children are in no way damaged by being vegans.

Notes

1 National Obesity Rates & Trends, NHANES, 2019. www.stateofobesity.org/obesity-rates-trends-overview/
2 Y.C. Wang, K. McPherson, T. Marsh, S.L. Gortmaker and M. Brown, "Health and economic burden of the projected obesity trends in the USA and the UK." *Lancet*, 378(9793), 2011. doi:815–82521872750
3 American Heart Association, "Nearly half of all adult Americans have cardiovascular disease." *ScienceDaily*. www.sciencedaily.com/releases/2019/01/190131084238.htm
4 Centers for Disease Control and Prevention (CDC), "Calorie consumption on the rise in United States, particularly among women." 2014. www.cdc.gov/nchs/pressroom/04news/calorie.htm
5 J. Ducharme, "About 90% of Americans don't eat enough fruits and vegetables." *Time*, 2017. https://time.com/5029164/fruit-vegetable-diet
6 H. Ritchie, "Which countries eat the most meat?" *BBC News*, 2019. www.bbc.com/news/health-47057341
7 B. Swinburn, G. Sacks and E. Ravussin, "Increased food energy supply is more than sufficient to explain the US epidemic of obesity." *The American Journal of Clinical Nutrition*, 90(6), 2009: 1453–1456 https://doi.org/10.3945/ajcn.2009.28595
8 A. J. German et al. "Dangerous trends in pet obesity." *The Veterinary Record*, 182(1), 2018: 25. doi:10.1136/vr.k2
9 H. A. Raynor and L. H. Epstein, (2001). "Dietary variety, energy regulation, and obesity." *Psychological Bulletin*, 127(3): 325–341. http://dx.doi.org/10.1037/0033-2909.127.3.325
10 D. F. Birt, T., Boylston, S. Hendrich, J.-L. Jane, J. Hollis, L. Li. J. McClelland, S. Moore G. J. Phillips, M. Rowling, K. Schalinske, M. P. Scott, and E. M. Whitley, "Resistant starch: Promise for improving human health." *Advances in Nutrition*, 4(6), 2013: 587–601, https://doi.org/10.3945/an.113.004325
11 N. Darmon, M. Darmon, M. Maillot and A. Drewnowski, "A nutrient density standard for vegetables and fruits: Nutrients per calorie and nutrients per unit cost." *J Am Diet Assoc*, 105, 2005: 1881–1887.

1 The ethics of veganism

The ineffectiveness of moral theories

Over the years, moral philosophers have offered many ethical justifications for vegetarianism and veganism. In fact, there are too many. There are so many that even philosophers are confused. Non-philosophers are even more confused. The issue of eating or not eating animals is so complicated that some moral experts are not sure what to think. Oxford moral philosopher Jeff McMahan still doesn't know if it's wrong to eat animals. In 2017, Olivia Goldhill interviewed Oxford moral philosopher Jeff McMahan, who is critical of the suffering of animals caused by intensive animal agriculture, and asked him whether he was against eating animals that are treated humanely, to which he answered, "That's what I'm trying to figure out."[1] People rely on bits and pieces of information that they read online or in magazines or hear in lectures and talks. The most popular arguments against using animals for food are based on different strands of utilitarian principles or the ethic of rights. These arguments have taken different forms but, basically, they are based on these two ethical approaches. Utilitarian arguments are based on the writings of Peter Singer,[2] while arguments about the inherent rights of animals are based on the writings of Tom Regan.[3] Such arguments say that we should be or become vegans because animals suffer and suffering is not desirable or it is not right. In this chapter, I discuss some of the moral theories that have been used to justify ethical vegetarianism and veganism. My goal is to show that these moral theories might excite philosophers, but they are unconvincing and incapable of moving people toward veganism. I suggest embracing bare virtues, that is, virtues, such as moderation and nonviolence, without the theoretical baggage of virtue ethics.

Moral theories and veganism

Take utilitarianism. The argument goes something like this: everything we do, the choices we make, the actions we undertake, ultimately generates pleasure and or pain. No one in his or her right mind would want to experience pain or suffering. Everyone desires pleasure or happiness. Thus, the point of morality is to maximize pleasure/happiness and minimize pain/suffering. Now, from here it gets a bit complicated. However, to have a rough idea of utilitarian ethics, let's consider the following example. Many people believe that lying is wrong or that it is not what a noble or respectable person would do. Utilitarians, in a peculiar sense, agree with such a statement. However, they add the following. Suppose that you find yourself in a predicament where your lying can save the lives of many innocent people. It would seem that it is better to lie and save many lives than following the principle "do not lie" and lose many innocent lives. Utilitarians then reason this way: If your lying produces on balance more good that evil (more pleasure/happiness than pain/suffering), then lying in *those* circumstances is the right thing to do. Consider now something positive, such as helping an elderly person cross the street, or donate money to the poor. Just because such actions are positive, according to utilitarian ethics, it does not mean that they are the right actions. It is easy to imagine, for example, donating money to a person who will use it to drink himself to death or for some evil purpose. Or, while helping an elderly person cross the street, a bus might run over him and kill him. Thus, the ultimate measure of morality according to utilitarian ethics is the capacity of an action to achieve the greatest overall amount of good for the greatest number of beings. A typical objection is that no one can possibly know the future ramifications of our decisions. In other words, something that we envision going well may actually lead to catastrophic consequences.

The next step is to consider the subjects of morality. Since as we have seen the goal of morality according to utilitarian ethics is that our actions should produce the greatest overall amount of pleasure and the least amount of suffering, it follows that the subjects of interest are only those beings that are capable of experiencing pleasure or suffering. Thus the argument is the following:

1 Pleasure is good; pain/suffering is bad.
2 Pleasure ought to be promoted; pain/suffering ought to be avoided or suppressed.
3 Animals have the capacity to feel pain or pleasure (they are sentient).
4 The goal of morality is to act in such a way as to promote as much pleasure as possible and as little pain as possible for the greatest number of sentient beings.
5 Raising animals for food causes more pain than pleasure overall.
6 If we stop raising animals for food, the number of animals that exist will be smaller, and so will the amount of suffering.
7 Conclusion: We should stop raising animals for food and become vegetarians because being vegetarians produces more pleasure than pain for the greatest number of sentient beings.

The approach that utilitarians take with respect to animal ethics is unsatisfactory for many reasons. To begin with, utilitarian ethicists who write in support of animal welfare argue for vegetarianism or flexible veganism. That is, in principle, milk, eggs, cheese, honey, wool, leather, and occasionally meat from animals raised in certain conditions that do not cause direct pain to the animals are morally permissible. Peter Singer, for example, does not argue that it is necessarily a problem that we use animals for human purposes because he does not regard the killing of animals as inherently immoral. According to Singer, most animals are not self-aware and thus lack the capacity to even care that we use them. Moreover, farm animals lack a concept of continued self; in other words, such animals lack the notion that they will continue to exist in the future and have future plans. Consequently, for Singer it may be morally acceptable to consume animal products as long as we eat only animals who have been reared and killed humanely. In a 2006 interview in *The Vegan*, Singer states:

> [T]o avoid inflicting suffering on animals—not to mention the environmental costs of intensive animal production—we need to cut down drastically on the animal products we consume. But does that mean a vegan world? That's one solution, but not necessarily the only one. If it is the infliction of suffering that we are concerned about, rather than killing, then I can also imagine a world in which people mostly eat plant foods, but occasionally treat themselves to the luxury of free range eggs, or possibly even meat from animals who live good lives under conditions natural for their species, and are then humanely killed on the farm.[4]

It is causing gratuitous pain that is wrong. But the question is, why is killing animals for food gratuitous pain? The answer is that it is not always gratuitous pain. In fact, utilitarians condemn only industrial farming, and not all forms of animal agriculture. Thus, utilitarianism argues that we who live in affluent societies must be vegetarians. In other circumstances, killing and eating animals can be morally permissible. The purported advantage of utilitarianism is practicability. However, the utilitarian argument is confused and confusing. The approach to animal welfare that utilitarians propose, evidently, has had a remarkable academic impact. But in practical terms, how does utilitarianism help animals? It really doesn't. Remember that utilitarianism is not concerned about specific individuals; rather, it is concerned about maximizing utility. In other words, utilitarianism is an ethical system that emphasizes aggregate benefits, whether the benefit is overall satisfaction of preferences or happiness or pleasure. It argues that raising animals for food runs counter to the principles of utility because it produces suffering. One of the main problems with the utilitarian approach is that it opens the door to paradoxical and incongruous notions, such as the one of Mary Temple Grandin,[5] who, in the name of compassion and understanding of animal suffering, designed certain mechanisms intended to reduce suffering for animals being led to slaughter. In my view, utilitarianism offers an unsatisfactory moral approach to animal ethics. Worse, it is claimed to be an action-guiding theory when in reality it is a confusing theory. It is easy to see how consuming animal-based diets is consistent with utilitarianism.

Most utilitarians point out that eating animals is a form of speciesism, which is a form of discrimination among species. Singer's rejection of speciesism is not entirely based on utilitarian principles. As Lawrence Finsen and Susan Finsen have noted, Singer "presents an important objection to the current treatment of animals that is not based on a utilitarian calculation but expressed in terms of demanding that we avoid speciesism."[6] In fact, Singer suggests that we should avoid speciesism regardless of the consequences of our actions. But this is not a utilitarian principle. Animals have an interest in not suffering, but they do not have an interest in continuing their existence because they lack such a concept. Hence, as long as we are careful not to make them suffer, according to utilitarianism it is permissible to kill them and eat them. I once asked Singer whether this is his actual position. His answer was, "That is still my view, although I think that in almost all circumstances it is better to avoid eating them."[7] The last clause of the above statement is crucial. To say that "in almost all circumstances it is better to avoid eating them" is not part of the utilitarian argument. It seems that in order to work, utilitarianism requires extra help from non-utilitarian principles. There is, therefore, an inevitable tension with Singer's view because even speciesism can be morally acceptable. This is a confusing (and confused) aspect of Singer's theory, which renders it impracticable and uneasy to follow.

Animals and rights

Another important argument is based on the notion that animals have inherent rights, the most important of which is the right to life. This is an argument developed by Tom Regan. Regan's approach is an example of deontological ethics, though Regan proposes a strong animal rights position. All animals have a right to life, and consequently he argues,

> The fundamental wrong is the system that allows us to view animals as *our resources*, here for *us*—to be eaten, or surgically manipulated, or exploited for sport or money. Once we accept this view of animals—as our resources—the rest is as predictable as it is regrettable.[8]

In *The Case for Animal Rights*. Regan argues that all mammals such as cows, pigs, goats, (those mammals that are typically eaten by people) over a year of age have the same basic moral rights as humans. Regan presents his argument, as Mary Anne Warren points out,[9] in three stages. The first stage is to note that animals, such as those I just mentioned, are more than mere fleshy machines. Animals are sentient creatures, endowed with memory, emotions, desires, identity over time, and other important mental characteristics that are relevantly similar to those possessed by humans. Consequently, Regan believes that animals that possess such capacities are subjects-of-a-life. And since subjects-of-a-life can be harmed or benefitted, it follows that animals can be harmed or benefitted. Using them for food can harm them. Therefore, we should not use them for food or other practices that can harm them.

The second stage, however, is more problematic. Regan argues that subjects-of-a-life have inherent value. While Kant, as we have seen, argued that animals are mere means to our ends, Regan argues that since animals are subjects-of-a-life, they are ends-in-themselves. According to Regan, all animals have inherent value, that is, inherent value does not come in degrees. To say that some animals have more values than others, according to Regan, is to adopt a perfectionist moral view, which assigns different moral value on the basis of certain characteristics, e.g., intelligence, species, etc. We know from history that such a theory could justify many unjust positions, such as slavery, male domination, racism, etc. Thus, Regan argues that we must reject the perfectionist view and adopt a view according to which we divide all living things into two categories: those that have inherent value, which have the same basic rights as humans, and those that do not have inherent moral value, which have no moral right. Consequently, all subjects-of-a-life must be regarded as having rights.

The third stage is to argue that all beings that have an inherent moral value must be respected; they must not be treated as mere means to our ends. This implies that we have a direct prima facie duty to respect and avoid harming all subjects-of-a-life. It is these considerations that generate moral rights. All morally valuable beings have a right to life. And rights imply obligations. We have the duty to avoid harming others or treating them as means to our ends—and moreover we have the duty to avoid harm and help others that are endangered.

Regan places a great deal of importance on the distinction of normal mature mammals and other species. In his defense, perhaps Regan wanted to emphasize normal mammals for practical reasons: firstly, animal agriculture treats normal mammals horribly. Second, Regan might have thought that if we accept his argument, a great deal of injustice and animal suffering can be avoided since they are mainly caused by animal agriculture. Perhaps, Regan thought that if we accept his subject-of-a-life principle, we would eventually be kind and considerate to all forms of life and even the environment. However, his emphasis on normal mammal generates difficulties. For example, what makes a mammal more special than a reptile or a bird? And why does the animal have to be a normal mature one to have inherent value? I find it peculiar that Regan proposes that there should be any correlation between age and inherent value when his very theory argues that inherent moral values does not come in degrees. One possible answer is that whenever we are uncertain of whether a being is a subject-of-a-life, we may give it the benefit of the doubt. However, how exactly is it to be applied? As Warren notes,

> If we try to apply this principle to the entire range of doubtful cases, we will find ourselves with moral obligations which we cannot possibly fulfill. In many climates, it is virtually impossible to live without swatting mosquitoes and exterminating cockroaches, and not all of us can afford to hire someone to sweep the path before we walk, in order to make sure that we do not step on ants.[10]

Indeed, according to Regan's argument, since animals, like humans, have a life that can go well or can be frustrated, they all have the right to life and to be treated with respect. But it does not seem to follow from this that every life has the same value. It is true that animals have the same value as humans? Granted, Regan states a qualification: that animals have a life that, like humans, can go well or badly *for them*. If a squirrel has a good or bad life, it depends on squirrely factors. However, does it follow that squirrels and humans should have the same basic rights? The very concept of right, especially human rights, is not an uncontroversial one. Humans, if any, have rights by virtue of the fact that they have an understanding of rights and they can enforce them upon others. As Carl Cohen stated,

> this much is clear about rights in general: they are in every case claims, or potential claims, within a community of moral agents. Rights arise, and can be intelligently defended, only among beings who actually do, or can, make moral claims against one another.[11]

Regan's argument is an admirable position in theory. However, it is not conducive to respect of animals. For one thing, it is unconvincing because it is based on the assumption that because animals, like humans, have a life that can be good or bad *for them*, it follows that both humans and animals have the same basic rights. This does not seem to follow if we consider that talking about rights, as many have pointed out, makes sense in a context in which the parties can intelligently make sense and impose certain rights. Animals, obviously, would be excluded.

Some theoretical problems

Problem 1) We can raise animals in ways that reduce pain and harm to the environment without altering the number of animals that are slaughtered

Surely, it would be possible to reduce animal suffering and harm to the environment even further. Suppose for example that farm animals were given a drug or, better still, were genetically manipulated such that they do not feel pain at all. And suppose also that certain measures were implemented to avoid the negative impact on the environment. In that case, then it would seem that from a utilitarian standpoint raising animals for food would be morally permissible. True, there remains the question of the loss of utility caused by the negative impact on human health resulting from consumption of animal products. This concern alone however, would not generate a moral obligation to become vegetarians. Meat eaters would be justified in taking a chance on their own health. Moreover, as I argue in Chapter 2, assuming that it was possible to eliminate animal suffering, this would not be the end of the conversation; for, there remain other moral concerns to be accounted for, such as aesthetic value loss, gustatory self-deception, and violence.

12   *The ethics of veganism*

Problem 2) *Animals do not suffer the same way as humans suffer.*

Animals feel pain. In light of current research in cognitive ethology and neuroscience we know that animals are sentient creatures. Here are two strands of problem:

First: it is hard to think how to motivate people to actively reduce the suffering of animals. Many people just do not care. Some people change their minds and decide to go vegan or vegetarian as a result of their watching YouTube videos that show the degree of suffering inflicted on animals on factory farms and slaughterhouses, reading papers, or discussing the issue. But for many others, those videos and other forms of information are irrelevant. They may not be concerned about the suffering for many reasons. For example, religious people might believe that a soul is required to feel pain, animals do not have souls, and therefore animals do not feel pain. Others who do not necessarily subscribe to religious ideals might just be indifferent to the pain of farm animals because such animals are distant and unknown to them. After all, farm animals exist because humans domesticated them for food.

For some people, this means that such animals do not belong in their circle of care or duty. Yet others might point out that admirable or commendable as it might be, caring for others is legitimately extended to a restricted circle. In other words, why should we care about people we never met before and perhaps never will meet? Why should I care about criminals or people who do not live like me? Surely, individuals who make such claims may be concerned about others. However, they may point out that just because we have moral obligations toward all, it does not follow that we ought to have the same degree of respect for others—even less for animals. People may point out that it is reasonable to care about their neighbors; but why care so much for a person or people who live on the other side of the world? And if one is plausibly justified in not caring less for others who live far away than those who live closer, it is also plausible that one cares even less about animals. Why? Perhaps the immediate reason is that while people relate with other people in many ways (same species, similar desires, etc.) people are very different from animals.

Second: What is a sentient being? The definition of sentience denotes the capacity to feel and to have subjective conscious experience. Is this capacity enough to assign moral importance? Here is an argument to consider: observation reveals that sentience is not an all-or-nothing property; rather it comes in degree. Functional adult humans are considered by many to be the paradigm of sentience. As we move down to lower beings, the degree of sentience also diminishes. Apes, for example, being the closest relatives to humans, are very much aware of their existence, though they lack the higher cognitive functions characteristic of humans. Squirrels exhibit an even lower degree of sentience than apes and humans. Moving down the ladder we find farm animals. The argument then is that when humans feel pain or pleasure, they are also aware that they are feeling pain or pleasure. Conversely, animals lack subjective awareness of their mental states. Not all animals. Perhaps, many would accept that the great apes are self-aware. But this doesn't matter because humans, usually, do not raise apes, for food. Animals like cows, pigs, chickens, lambs, on the other hand, are sentient but, if they are in pain, (the argument goes) they are not aware that they are.[12] Consequently, many argue that it is morally permissible to kill and eat those animals because raising and killing them can cause them pain, but they are not aware of that pain.

Although there is a general agreement that farm animals are sentient beings among philosophers, there is a vast literature on exactly what it means that, and to what degree, animals feel pain and suffer. To make this matter even more complicated, there are philosophers who argue that animals like pigs, chickens, and cows can suffer. Michael Murray for example proposes a Neo-Cartesian account of animal suffering. He suggests that animal pain and suffering are based on neurological complexity, the greater the complexity the greater the awareness of pain.[13] Also, Clare Palmer thinks that many organisms may only be capable of "unconscious responses to pain."[14] Her argument is that "research on human fetuses indicates withdrawal reflexes before the development of the thalamo-cortical circuits associated with pain perception."[15] Murray argues that organisms exhibit pain behavior when exposed to pain-inducing stimuli. For example, protozoa, amoebae, etc., are not sentient to the degree that they have conscious experience of pain although they might recoil when poked with an object. Animals with a more neurologically complex structure experience first order pain, that is, they feel pain, but are not aware that they are feeling pain. Only humans, and perhaps other primates, experience second order pain, which is the awareness of suffering, i.e., they know, anticipate, and reflect upon their pain experience.

Thus, some argue that the point is not whether these creatures can feel pain, but rather whether they have an interest to avoid pain. Human infants arguably lack such interest, and yet nobody would suggest that gratuitous infliction of pain upon human infants is morally permissible. Most notably, Peter Singer has been arguing for years that the cognitive capacities of organisms make a great moral difference in the way we ought to treat them. Peter Singer's view is very famous (or maybe infamous), and thus I don't need to explain it in detail here. Suffices to say that Singer argues that most animals, as well as infants and fetuses, lack a sense of self as a continuing subject of experience. Consequently, it would not be wrong to terminate their lives, provided that there exist good reasons for doing so. Of course, one would have to dispute the premise that killing animals for food causes gratuitous pain. The result is that cognitive distinctions or distinction made on the basis of organisms' having an interest to avoid pain are highly controversial and ultimately not convincing one way or the other. Furthermore, some philosophers even argue that sentience does not stop at farm animals, but continues down to clams, insects, plants, and bacteria, and extends even further down than bacteria. In other words, it is the view that all true individuals are sentient—a position that Jan Deckers refers to as "pan-sentientism."[16]

Such notion of sentience is so controversial that some philosophers avoid factoring in the moral equation. For example, in his recent book, *Can Animals Be Persons?* Mark Rowlands does not attempt to give a solution to the question of whether animals have minds. Rather he offers a dissolution of the problem. We have a direct experience of other human minds and a complete certainty of this fact, he argues, and our empirical evidence of other human minds is more certain than our inferences. As Wittgenstein puts it, "The human body is the best picture of the human soul."[17] The direct perception view that Rowlands employs is a very interesting one and has three steps: First, there is a distinction between seeing and seeing *that*; second, there is a distinction between formal and functional descriptions of behavior; third, he argues that functional descriptions of behavior are disguised psychological descriptions. These three steps combined lead to the conclusion that "we can often see the mental states of animals."[18]

Distinctions between seeing and seeing that is a distinction between certain actions or phenomena that we do not directly see, but we see that they are certain particular events. For example, I do not see the summer season, but I see *that* it is summer. Rowlands gives an example of a tornado. Typically, one might say that he sees a tornado. To be precise, it is not the tornado itself that is being seen, but rather its effects, rotating objects, dirt, water, etc. Regarding behavior, it is not always possible to tell the nature of a behavior simply by looking at it; nevertheless, the behavior is visible. Thus, a behavior can be functionally described in psychological terms.

Rowlands explains the behavior of animals, arguing that functional descriptions of their behaviors are none other than a "disguised mental or psychological descriptions."[19] It is the way we take the world to be that explains our behavior. In other words, the functional descriptions of the behavior of animals are not psychologically neutral; rather, they reveal cognitive attitudes of a being, in this case animals. In other words, using Rowlands' example, when a dog performs a play bow to initiate play, we cannot see *that* the dog is initiating play but we can see the dog doing it—that is, I see psychological states of the dog. As Rowlands puts it, then, "If we want any sort of illuminating science of animal behavior, we should acknowledge that our primary access to the minds of animals is not through inference but through perception."[20] This Rowlands makes clear, is not a solution to other human or animal minds, but rather dissolution. Thus, it may still be objected that while humans are conscious, animals are not.

Problem 3) We can grow meat and more in labs (I discuss this in Chapter 3).

If suffering and the environment are the only concerns, synthesizing animal products seems to be a viable solution. In fact, producing meat in labs could reduce the ecological footprint left by the farm animals' sector and considerably reduce the amount of suffering. Furthermore, in addition to meat, now scientists are beginning to talk about in vitro milk and eggs.[21]

Therefore, having considered the foregoing problems, it is evident that the traditional theories in animal ethics do not make a convincing case to support the view that animals should not be considered as food and property. Perhaps one of the most difficult obstacles in the discussion of our relationship with animals is that vegan philosophers try to convince people that we should eat vegetables instead of animals by using ethical theory. But how effective is this tactic? This is the topic considered in the next section.

Ethical theories are for philosophers not for people

The typical way ethicists address moral question is by starting from a favored ethical theory and moving to a certain moral conclusion or prescription, as we have seen in the previous discussion. Utilitarianism, rights ethics, and deontology are procedural theories. This means that they offer a procedure to determine the right action. Employing procedural ethical theories for most people (including for some moral philosophers) who face moral conundrums is highly impractical. First, theories such as utilitarianism or deontology are very complicated for most people. Second, such theories require that the agent accept a nexus of theoretical assumptions without which a theory would be useless. Conversely, virtue ethics has the advantage of working even if it is bare, i.e., devoid of its theoretical principles because after having stripped it off of its conceptual framework, what remains is a robust notion of the virtues. As I will suggest, the notion of a virtue is very practicable and more helpful than procedural ethical theories in deciding what to do.

Procedural moral theories and bare virtue ethics

I have been teaching various philosophy courses for many years now, especially courses in ethics and applied ethics. I always wondered whether these courses are effective in some way. More specifically, I often wonder whether learning ethical theories helps make better moral decisions. The typical way ethicists address moral question is by starting from a favored ethical theory and move to a certain moral conclusion or prescription. For example, in "Utilitarianism and Vegetarianism" Peter Singer writes, "I'm a vegetarian because I am a utilitarian."[22] Here Singer makes clear that this is the correct way to proceed. Namely, first you choose a moral theory and then you follow where it leads. In the case of Singer, utilitarianism has led him to vegetarianism. Singer notes that it would be improper to approach moral questions the other way around, i.e., from one's intuition to the moral theory that matches. He refers to this approach as "a curious inversion."[23] He writes,

> Some philosophers think that the aim of moral theory is to systematize our common moral intuitions. As scientific theories must match the observed data, they say, so must ethical theories match the data of our settled moral convictions. I have elsewhere argued against the inbuilt conservatism of

this approach to ethics, an approach which is liable to take relics of our cultural history as the touchstone of morality.[24]

The approach that starts with a preferred ethical theory to answer moral questions relies on the implicit notion that learning ethical theories is necessary to navigate through ethical problems. As Dien Ho points out, "The idea that we need ethical theories to tell us what we ought to do might strike most laypersons as awkward and artificial; e.g., consider how odd it sounds to decide whether one ought to continue a pregnancy by seeing if it maximizes utility."[25] Indeed, it is awkward to address a moral issue by consulting moral theories and following a theory's principles. I am not sure that even ethicists do it, let alone non-philosophers. Most people do not understand moral theories or do not find them particularly helpful. Furthermore, most moral theories are too complex and highly disputed and criticized. One might say that complex or not, ultimately one should embrace the most logically consistent theory, the one that has in its favor the best arguments. However, deciding which arguments are the most plausible is also a matter of intuition. As Rob Lawlor suggests,

> Moral theories should not be discussed extensively when teaching applied ethics … students are either presented with a large amount of information regarding the various subtle distinctions and the nuances of the theory and, as a result, the students simply fail to take it in or, alternatively, the students are presented with a simplified caricature of the theory, in which case the students may understand the information they are given, but what they have understood is of little or no value because it is merely a caricature of a theory.[26]

Many philosophers today are puzzled about normative theory in the sense that they cannot tell which theory is correct. As a result, many moral philosophers no longer insist like they did in the past that to resolve moral problems it is necessary first to identify a moral theory. In his introduction to ethics textbook Russ Shafer-Landau aptly encapsulates what I am trying to say:

> [M]ost philosophers who grapple with real-life moral problems begin not with the grand normative theories but rather with more concrete principles (keep your word, don't violate patient confidentiality, avoid imposing unnecessary pain, etc.) or examples that we are already expected to accept. Armed with these, the authors then try to show that a commitment to them implies some specific view about the problem at hand.[27]

Another practical example is Oxford philosopher Jeff McMahan, who recently expressed reservations about his moral conclusions. Similarly, Brad Hooker makes the point that we have much more confidence in our judgments about which particular *pro tanto* duties we have than we have in our judgments about which moral theory is correct.[28]

I do not intend to undermine the importance of teaching ethics and learning about moral theories. Rather, I wish to point out that academic philosophers discuss ethics and publish highly sophisticated papers abut ethics but seldom contemplate the application of their high-minded work. If we want to make a difference, we must take stock of the way in which we teach and approach morality. My ambitious thesis is that virtue ethics is the most useful moral approach to applied ethics because unlike other moral theories (especially deontology and utilitarianism) it helps resolve moral problems even if its theoretical basis are dropped.

Virtue ethics

First off, consider some important theoretical basis of virtue ethics. Virtue ethics is a moral approach that differs significantly from utilitarianism and deontology. Surely, deontology differs significantly from utilitarianism. However, both theories are procedural in that they purport to give moral guidance, and show which acts are morally permissible or obligatory and which acts are not. They also offer a standard of rightness because they provide the conditions that make actions morally right. Thus, by saying that virtue ethics is a moral approach that differs from deontology and consequentialism, I mean that the primary aim of virtue ethics is not to be a decision procedure. Arguably, this aspect of virtue ethics, among many others, is what turned many thinkers away from it for a long time, though it is the very aspect that has drawn many others to it again.[29] Virtue ethics begins with a standard of good, but does not claim, at least directly or by way of some sort of calculation, to guide our actions.[30] As many introductions to Aristotelian ethics point out, virtue ethics is concerned with determining how to be a good person. With regard to being a standard of rightness, virtue ethics does provide the condition that makes actions right; but even this aspect differs from consequentialism and deontology. For example, Rosalind Hursthouse discusses this issue and proposes the (V) rule: "An action is right if it is what a virtuous agent would characteristically do in the circumstances."[31]

Virtue ethics says that a right action is an action among those available that a perfectly virtuous human being would characteristically do under the circumstances. By a perfectly virtuous person it is meant an individual having admirable character traits such as moderation, fairness, courage, honesty, generosity, civility, friendliness, and wittiness.[32] The virtues are reliable character traits that guide one's values, emotions, attitudes, and desires according to reason. A virtuous person, for example, will avoid overeating or eating just for pleasure any amount and type of food; will not take any bribe; will repay a debt; will not be party of an unjust cause; will help others in need if possible, and more. Suppose I find myself in a moral predicament. If I embrace the principles of virtue ethics to figure out how to resolve the issue I may appeal to the very concept of the virtue by asking, What is moderate? What is consistent with nonviolence? Some philosophers worry that it might be complicated for people who live in a fascist societies to figure out what is just or moderate. However, most societies are democracies where people have a good idea of what justice and fairness and moderation are. Also, even in those societies, people can come to similar conclusions by using reason.

Another way to find out what one ought to do in a particular circumstance is to consult a moral exemplar or one who knows a great deal about moral issues. In Christianity, for example, Jesus is the moral exemplar. Christians often ask, "What would Jesus do?" in order to determine the right way to live and to act. In other words, as Jason Kawall points out, "we can use virtuous actions as a heuristic in identifying virtuous agents, but we can use these actions in this way precisely because they are the sorts of actions that virtuous agents would perform."[33] The primacy of virtue does not require any metaphysical commitment to the nature or virtues. Rather, it involves appealing to those character traits that are noble and good-making and beneficial to society. And I do not mean this in a utilitarian sense. I mean it in the sense that virtue ethics, properly understood, advocate altruism and social harmony. As Julia Annas states, "We care about being generous, courageous, and fair. This looks as though we care about other people, since what we care about is having a disposition to help others, respect their rights, and intervene when they are threatened."[34] These are traits that we would value in people, in light of what is rational and conducive to a noble life. Again, we can accept these traits as virtues without assuming eudaimonia, the notion of human function, or even morally right actions.

Virtues should be understood as triads. Aristotle argues that a virtue lay in the middle of two extremes, one of excess the other of defect: "the mean by reference to two vices: the one of excess and the other of deficiency"[35] Courage—for example, lies between foolhardiness and cowardice; compassion lies between callousness and indulgence; temperance is a mean between the excess of intemperance and the deficiency of insensibility. Too much and too little are always wrong.[36] Aristotle's ethical doctrine is clear: avoid extremes of all sorts and seek moderation in all things. A person performs virtuous action only if 1) he knows that the action is virtuous, 2) he chooses to do the action for the sake of being virtuous, and 3) his action proceeds from a firm and unchangeable character. In short, an action is truly virtuous if it is such as a virtuous person would do.[37]

Nowadays, there are many interpretations of virtue ethics, that is, many virtue theories, which differ from Aristotle's original formulation. The modern world is socially and politically more complex than the Greek polis. Consequently, many contemporary virtue ethicists have interpreted Aristotle's principles in ways that make sense to the present. This however does not mean that modernity has made the virtues redundant. As Gong and Zhang rightly note, "Modern ethical life is still the ethical life of individuals whose self-identity contains the identity of moral spirit, and virtues have a very important influence on the self-identical moral characters."[38] There are several modern interpretations of virtue ethics.[39] As Rosalind Hursthouse notes, the proponents of virtue ethics nowadays "allow themselves to regard Aristotle as just plain wrong on slaves and women." We also regard other traits, such as charity, benevolence, and others, to be on the list of virtues, despite Aristotle's failure to do so.[40] In my view, the principal aspect of virtue ethics is the virtues themselves and acting from them.

Virtues and vices

Standard interpretations of the *Ethics* usually have Aristotle emphasize the role of habit in conduct. That is, virtues, according to Book II, chapter 4, are identified as *hexis; hexis* are habits that are conducive to the good life. Besides habit, or custom, *hexis* also denotes an active condition, a state that manifests itself in action. Moral virtues arguably are the most important aspect of virtue ethics. Virtues are admirable character traits, or desirable dispositions, which contribute, among other things, to social harmony. These character traits enable us to act in accordance with reason. Virtues enable us to feel appropriately and to have the right intention and feelings in a given situation. The person whose character is not virtuous may do what appears from the outside to be the right thing to do, but his motives will leave something to be desired. For example, a person who has acquired the virtue of honesty will usually tell the truth. Telling the truth for the right reason at the right time in the right situation is what an honest individual would do, and not because one fears the negative consequences of being found out for telling a lie.

Virtually all philosophers nowadays recognize a number of virtues.[41] These are the characteristics that we desire for ourselves, our friends, our neighbors, our children—in fact, for all people. These are moral characteristics that those who are serious about morality strive to attain. Cruelty, viciousness, and malice are anti-social and thus are not moral traits conducive to a good life. Our moral experiences happen socially. In a society, characteristics such as magnanimity, nonviolence, benevolence, moderation, and all other character traits that have a good-making disposition are desirable for their own sake. Granted, circumstances might require, say, that a parent lie to her child telling him that Santa will bring a gift or that grandma is in a better place now, or for a ruler to lie to his citizens because telling the truth might lead to conflict. But the motivation for lying proceeds from certain virtues that possess a good-making disposition that aims at the good life. For example, the parent's lying is motivated by her love for her child, knowing that telling the truth—Santa Claus is not real, kid!—may break his heart. So, the virtues are good (in the sense that they are characterized by benevolent intentions) moral characteristics required by humans to achieve happiness.

Therefore, at a very practical level, the virtues are quite clear and basic concepts, which are shared among different peoples. It is understood the distinction between one who lies from a noble disposition of character–for example one who tells her aged, dying parent that he will recover—and another who lies for a malicious, malevolent or similar purpose—one who, for example, lies to get out of trouble or with a selfish or destructive motive. When we say that one is a chronic liar, we do not use the term arbitrarily. Rather, we say that the liar's character is not virtuous or that it is ignoble. And we recognize that an individual who lives his life by always telling the truth (except when telling the truth might hurt somebody or a lie is required for a good cause) is a trustworthy individual and socially valuable. Or, in other words, he is a virtuous individual or has virtue.

The purpose and meaning of life

Aristotle argued that the purpose of human existence is to achieve a state of *eudaimonia*, which is a difficult term to translate. Generally, *eudaimonia* is intended as the sort of contentment or satisfaction that is deep, lasting, and worth having. It is the sort of happiness that one experience in hindsight when contemplating life as a whole. It is the sort of happiness that makes one be proud and satisfied with one's life. The closest approximation is "flourishing." Think of the analogy of a bud that under ideal circumstances grows into a flower. Naturally Aristotle does not intend flourishing as a metaphor. Aristotle believes that a human being flourishes as a result of leading a good life having fulfilled the purpose and function of human beings. According to Aristotle, all human beings from all walks of life, whether they like it or not, have a function and purpose. The function is to use reason and the purpose is to acquire the virtues in order to flourish. As Philippa Foot puts it:

> Men and women need to be industrious and tenacious of purpose not only so as to be able to house, clothe and feed themselves, but also to pursue human ends having to do with love and friendship. They need the ability to form family ties, friendships and special relations with neighbors. They also need codes of conduct. And how could they have all these things without virtues such as loyalty, fairness, kindness and in certain circumstances obedience?[42]

In other words, virtue ethics represents a holistic approach to morality because it focuses on a deep understanding of our social lives and existence.

Practical wisdom

Practical wisdom is the idea that a virtuous agent, besides the virtues, must acquire an intellectual skill that enables him to make the right decisions when faced with particular moral issue. Aristotle discusses practical wisdom in Book VI of the *Ethics*. Practical wisdom (*phronesis*) is an intellectual virtue. Aristotle distinguishes two kinds of reasoning, theoretical reason and practical reason.[43] Theoretical reason investigates the unchangeable and aims at the truth. Practical reason investigates things that are subject to change and aims at making good choices. To make good choices, not only must our reasoning be correct, but we must also have the right desires.[44] The person with practical wisdom deliberates well about how to live a good life.[45] So practical wisdom is "a true and reasoned state of capacity to act with regard to the things that are good or bad for man."[46]

Thus, VE concerns itself with virtue as essential to achieve a good life. The virtues are character traits and practical intellect that are necessary for happiness. As Peter Geach famously pointed out, "Men need virtues as bees need stings."[47] The virtues give the right ends in the fulfillment of life, and thus function as internal guides. For example, generosity makes us help others in need, and practical wisdom enables us to understand whether others are in real need of our help and moreover how to go about helping them.[48]

Bare virtue ethics

Virtue ethics has been fairly thoroughly praised and criticized by many philosophers. One ongoing complaint is the very concept of virtue. Many moral psychologists, for example, argue that there are no such things as virtues and that people exhibit off-and-on episodes of virtuous behavior that vary depending on the person, her mood, the weather, the side of the bed on which she woke up, and many other factors. Furthermore, they say, virtuous behavior is not predictable. Ross and Nisbett, for example, argue that "one cannot predict with accuracy how particular people will respond ... using information about an individual's personal dispositions."[49] Based on empirical evidence, many philosophers[50] doubt that virtues exist or that the so-called virtues can predict and explain people's behavior.[51] These criticisms, however, miss the target. As Miguel Alzola points out, "a behavioral disposition does not constitute a virtue. At a minimum, the agent must have a good motive for her behavior if her disposition is to count as a virtue."[52] In other words, such empirical evidence is based on people's behavior and behavior's predictability and tries to attribute virtue to particular conducts. As Alzola points out, is that one can be virtuous regardless of the behavior that is typically expected from the possession of a certain virtue. For example, a person may be generous without being ostentatious about it, while another person may be quite lavish in donating goods to others and helping others without the appropriate internal feeling and disposition for generosity. Thus, as Aristotle explains, a virtue is a trait of character that is manifested in habitual actions. For example, one does not possess the virtue of temperance (moderation) because she approaches life in a moderate way only occasionally or whenever it is to her own advantage. Furthermore, a temperate person is not one who struggles to contain herself. Rather it is one to whom moderation (for example, eating or drinking moderately) comes quite naturally and feels good. The temperate person, some would say, is moderate as a matter of principle, though I prefer to say that the temperate person (or the courageous person, the magnanimous person, the just person, etc.) is moderate as a matter of character because his actions, as Aristotle puts it, "spring from a firm and unchangeable character."[53]

This description of virtue, however, does not distinguish virtues from vices. After all, vices are also traits of character manifested in habitual action. Edmund L. Pincoffs, suggests virtues and vices are qualities that we prefer or avoid: "Some sorts of persons we prefer; others we avoid.... The properties on our list [of virtues and vices] can serve as reasons for preference or avoidance."[54] Regarding various criticisms, Alzola states, "None of this, however, denies that character traits of the sort postulated as virtues do exist. None of this impairs our capacity to become morally better persons and attain the excellences of character that virtue ethicists call the virtues."[55]

Thus, in different societies, each virtue may be interpreted differently, and different sorts of actions are regarded as satisfying each of them. Since people lead particular lives and experience particular sorts of circumstances, they will have the need to express some virtues more than others. However, a virtue is not merely a matter of social convention. The virtues are not assigned by social convention but by basic facts that are common to all humans.

I want to suggest that many of our current moral issues can be better understood through the lenses of virtue ethics. When I say that virtue ethics is helpful, I mean that it is more helpful than deontology and consequentialism. These moral theories, as elaborate as they are, require a moral agent an impossible task, commitment to obscure principles, and ultimately treat morality in a rather distant way. We have heard this before: Kant's morality seems to be too concerned about rules and not very much concerned about feelings emotions and virtue. Utilitarianism as well treats morality as a matter focused on the amount of pleasure or suffering that sentient that individuals experience. What does this mean in practical terms? Let's take the instant case of vegetarianism or veganism. According to utilitarianism (Singer) we should act toward the environment in such a way that our actions bring about as much utility as possible. When we think about what we do to animals, according to singer it is clear that vegetarianism will bring about the greatest good for the greatest number of sentient beings.[56]

Within utilitarianism people, animals, plants, and nature, are regarded not as having an intrinsic value, but rather as "things" that can contribute to the satisfaction of aggregate preference/pleasure/happiness. The *only* objective value for utilitarianism is pleasure and the *only* objective disvalue pain. Thus, utilitarians are not exactly concerned about sentient beings as such; rather, utilitarians are concerned with the amount of utility that such beings can contribute to the greatest number. I don't want to appear too simplistic here. Naturally there are many details missing. But here the point is not to assess utilitarian arguments, but rather determine the effectiveness and practicability of utilitarianism. At any rate, utilitarianism is not very complicated: If they feel pain, they count in a utilitarian calculus. If they don't feel pain our actions toward them must be measured on the basis of the way in which our actions will frustrate or promote the maximization of utility. With regard to deontology, since it is a theory that emphasizes following rules, it is not apparent to most moral individuals which rules one ought to follow and, moreover, from where such putative rules are derived.

Here is the problem: Consider deontology. The only possible way to profit from such a theory is to accept the notion of universality of moral rules. Utilitarianism, similarly, requires accepting the notion of some sort of utilitarian calculus, which in its turns requires accepting the notion that the sole value is pleasure/happiness and the sole disvalue is pain/unhappiness. In sum, a rights theory, for example, says that we ought to do so and so on the basis of certain inherent rights. What if I don't believe in inherent rights? Utilitarianism says that we ought to act in such and such a way on the basis of what is conducive to the greatest good for the greatest number—where the greatest good is aggregate pleasure/happiness/satisfaction of preferences. What if I do not embrace that principle on the ground that I do not know what the future is like and do not hold that the right thing to do is measured on the basis of how my actions lead to the maximization of utility? What if I do not accept that pleasure/happiness/satisfaction of preferences is the only objective good? Deontology says that I ought to act in such and such a way because reason shows that I ought to follow a categorical rule. What if I don't believe that there are categorical rules? What if I don't like the idea of acting without any regard for the consequences? If I strip off the main principles and assumptions of these theories, I am left empty handed. These theories, to work, to be effective, require that embracing and accepting their respective principle. Such theories rely on a complex nexus of inseparable principles. Without their respective principles, those theories are useless. Just consider any moral predicament. Consider making use of deontology or utilitarianism to tackle it. The first major issue is that such theories are quite intricate. An average individual involved in some moral predicament would find any of those theories more complicated than the very issue at hand. The second issue is that each of these theories are supposed to work on the basis of a delicate and complex nexus of principles: Deontology—universalizability principle; the role and significance of rationality; the notion of categorical rules; perfect and imperfect duty, and more. Utilitarianism—utility; maximization of utility; disutility; rule vs. act utilitarianism; the notion that happiness/pleasure/satisfaction of preference is the only objective good; quality/quantity of happiness/pleasure/satisfaction of preferences, and more.

McDowell,[57] among others, maintains that it is impossible to "codify" moral judgment. He writes:

> If one attempted to reduce one's conception of what virtue requires to a set of rules, then, however subtle and thoughtful one was in drawing up the code, cases would inevitably turn up in which a mechanical application of the rules would strike one as wrong—and not necessarily because one had changed one's mind; rather one's mind on the matter was not susceptible of capture in any universal formula.[58]

Also, Daryl Koehn suggests that reliance on principles is an attempt to find a "mechanical algorithm for making decisions."[59] And Jonathan Dancy argues that,

> [W]e can perfectly well rely on people by and large to do what is right in the circumstances. We don't need principles to tell them what to do, or to determine what is right, or to tell us what they are likely to do, any more than we need principles of rationality to be in place before we can begin to rely on people by and large to act sensibly.[60]

In other words, it is evident to many ethicists that the way in which many of the procedural moral theories (especially those that propose calculations and categoricals, such as deontology and consequentialism) suggest approaching moral issues is just not the way most people do or could do ethics. As Robert C. Solomon states, "we don't actually do ethics that way."[61]

Considering deontology, it is undeniable that Kant's logical conclusion about the morality of lying is that lying is morally wrong, even when a murderer comes to our door asking about the whereabouts of a friend of ours.[62] It is true that many Kantian scholars have tried to show that Kant's murderer example does not undermine deontology because it does not mean that we must blindly follow all principles relevant to a situation. But then, what are we supposed to follow if not our feelings, reason, and common sense? It is clear that in such a situation any good person—and by "good" I mean one who possesses certain virtues—exercising judgment would not follow universal rules, but asks herself, "What is the just, compassionate, moderate, courageous thing to do here?" Many defenders of Kant's moral thought have indicated that in such a situation, the principle of helping others overrules the principle that we may never lie. So, one could consistently refuse to answer, tell a half-truth, or even tell the murderer a total lie. Perhaps this is indeed the way in which Immanuel Kant envisioned his theory. However, it is clear that most people would make such a judgment independently of Kant's elaborate moral system. What I propose is that an individual faced with such a predicament would in fact resort to very basic notions of virtue. These notions are virtually universal and require no specific metaphysical or epistemological commitment to particular principles or procedures. It only requires appealing to virtue itself. This is what I call bare virtue ethics.

Conclusion

To be clear once again, my discussion here is not intended as a rejection of ethical theory. I have discussed the issue of helpfulness in adopting and following moral theories. What I have suggested is that most people—in fact, even most ethicists themselves—when faced with moral conundrums can hardly follow the procedures described by procedural moral theories, such as utilitarianism or deontology. I suggested two main reasons: One is that these theories are highly controversial and complicated, and this is not news to moral philosophers; two is that such theories rely on a nexus of principles that require total commitment in order to be practicable. On the other hand, a virtue-oriented attitude is more profitable than all moral theories because it works even after it is stripped off of all theoretical principles, such as *eudaimonia*, function, etc.

In the next chapter, I will offer a practical example of bare virtue that relies on aesthetic and integrity values. First, rearing animals and transforming them into food are aesthetically repugnant practices. Slaughtering billions of animals adds into the world a great deal of un-aesthetic values sufficient to condemn not only animal agriculture, but also non-industrial farming. This argument has two strands: the avoidance of unnecessary ugliness in the world in the form of slaughtering, blood, and the degradation of the environment, and the avoidance of unnecessary violence. Since we should value non-violence as a virtue, and the practices of using animals for food are inherently and unnecessarily violent, it follows that we should avoid using animals for food.

Notes

1 Olivia Goldhill. "Scientists say your 'mind' isn't confined to your brain, or even your body." *Quartz*, 2016. https://qz.com/1102616/an-oxford-philosophers-moral-crisis-can-help-us-learn-to-question-our-instincts/
2 See Peter Singer's *Animal Liberation*. HarperCollins, 1975.
3 Regan, Tom. *The Case for Animal Rights*. Berkeley, CA: University of California Press, 1983.
4 Rosamund Raha, "Animal liberation: An interview with Professor Peter Singer," *The Vegan*, Autumn 2006: 19.
5 See Mary Temple Grandin, www.templegrandin.com
6 Lawrence Finsen and Susan Finsen, *The Animal Rights Movement in America: From Compassion to Respect*, New York: Twayne Publishers, 1994: 186.
7 Personal e-mail to Peter Singer, March 29, 2016.
8 In Peter Singer (ed.), *In Defense of Animals*, New York: Basil Blackwell, 1985: 13.
9 Mary Anne Warren, "Difficulties with the strong animal rights position," *Between the Species*, 2(4), 1987: 434.
10 Ibid.: 436.
11 Carl Cohen, "The case for the use of animals in biomedical research." *The New England Journal of Medicine*, 315, 1986: 865.
12 See Michael Murray, *Nature Red in Tooth and Claw: Theism and the Problem of Animal Suffering*, Oxford University Press, 2008.
13 Ibid.
14 Clare Palmer, *Animal Ethics in Context*, Columbia University Press, 2010: 14, 18.
15 Ibid.: 12.
16 Jan Deckers, *Animal (De)liberation: Should the Consumption of Animal Products Be Banned*, London: Ubiquity Press, 2016: 70.
17 Wittgenstein, *Philosophical Investigations*, Part II: 178.
18 Mark Rowlands, *Can Animals Be Persons?* Oxford University Press, 2019: 38.
19 Ibid.: 41.
20 Ibid.: 46.
21 Mouat, M. J. and Prince, R. "Cultured meat and cowless milk: On making markets for animal-free food." *J. Cultural Econ.*, 11, 2018: 315–329. doi:10.1080/17530350.2018.1452277
22 Singer, Utilitarianism and vegetarianism, *Philosophy & Public Affairs*, 9(4), 1980: 325.
23 Ibid.: 326.
24 Ibid.: 326.
25 Dien Ho, "Making ethical progress without ethical theories." *AMA Journal of Ethics*, 17(4), 2015: 290.

26 R. Lawlor, "Moral theories in teaching applied ethics." *Journal of Medical Ethics*, 33 (6), 2007: 370.
27 Russ Shafer-Landau, *The Ethical Life: Fundamental Readings in Ethics and Contemporary Moral Problems*, 4th edn. Oxford University Press, 2018: 7.
28 Brad Hooker, Intuitions and moral theorizing. In Stratton-Lake, P. (ed.) *Ethical Intuitions and Moral Theorizing*. Oxford: Oxford University Press, 2002: 182.
29 For a discussion on the decline and revival of virtue ethics see Daniel C. Russell's "Virtue ethics in modern moral philosophy," and Dorothea Frede's "The historic decline of virtue ethics" in *The Cambridge Companion to Virtue Ethics*, Cambridge University Press, 2013.
30 Aristotle, 2002: 1109a 21f.
31 Rosalind Hursthouse, *On Virtue Ethics*. Oxford: Oxford University Press, 1999: 28.
32 Aristotle, 2002: III. 6–v.
33 J. Kawall, "In defense of the primacy of the virtues." *Journal of Ethics and Social Philosophy*, 3(2), 2009: 4.
34 Julia Annas, "Virtue ethics and the charge of egoism." In P. Bloomfield (ed.) *Morality and Self-Interest*. New York: Oxford University Press, 2007: 205.
35 Aristotle, 2002: II. 5.
36 Aristotle, 2002: II. 6.
37 Aristotle, III. 1.
38 Q. Gong and L. Zhang, "Virtue ethics and modern society – A response to the thesis of the modern predicament of virtue ethics." *Frontiers of Philosophy in China*, 5 (2), 2010: 255.
39 C. Swanton, "The definition of virtue ethics." In Russell, Daniel C. (ed.) *The Cambridge Companion to Virtue Ethics*. Cambridge University Press, 2013.
40 Rosalind Hursthouse, *On Virtue Ethics*. Oxford: Oxford University Press, 1991: 8.
41 A. Comte-Sponville, *A Small Treatise on the Great Virtues: The Uses of Philosophy in Everyday Life*. Metropolitan Books, 2001.
42 Philippa Foot, *Natural Goodness*. Oxford University Press, 2001: 44.
43 Aristotle, VI. 1.
44 Aristotle, VI. 2.
45 Aristotle, VI. 5.
46 Aristotle, VI. 5.
47 Peter Geach, *The Virtues*. Cambridge University Press, 1977: 17.
48 Aristotle, VI.I.2.
49 L. Ross and R. E. Nisbett, *The Person and the Situation: Perspectives of Social Psychology*. Philadelphia: Temple University Press, 1991: 2.
50 G. Harman, "The nonexistence of character traits." *Proceedings of the Aristotelian Society*, 100, 2000: 223–226
51 See for example John M. Doris, *Lack of Character: Personality and Moral Behavior*. Cambridge University Press, 2002.
52 Miguel Alzola, "The possibility of virtue." *Business Ethics Quarterly*, 22(2), 2012: 387
53 Aristotle, 1105a-b.
54 E. L. Pincoffs, "Two cheers for Meno: The definition of the virtues." In Shelp E. E. (ed.) *Virtue and Medicine. Philosophy and Medicine*, vol. 17. Dordrecht: Springer, 1985: 114.
55 Alzola, 2012: 396.
56 Peter Singer, "Utilitarianism and vegetarianism." *Philosophy & Public Affairs*, 9(4), 1980: 325–337.
57 John McDowell, "Virtue and reason." *The Monist*, 62, 1979: 331–350; John McDowell, "Two sorts of naturalism." In Altham, J. and Harrison, R. (eds), *Virtues and Reasons: Phillipa Foot and Moral Theory*, 1995: 149–179. New York: Oxford University Press.
58 McDowell, 1979: 336.

59 D. Koehn, "A role for virtue ethics in the analysis of business practice." *Business Ethics Quarterly*, 5, 1995: 534.
60 J. Dancy, *Ethics without Principles*. New York: Oxford University Press, 2004: 133.
61 R. Solomon, *Ethics and Excellence: Cooperation and Integrity in Business*. New York: Oxford University Press, 1992: 114.
62 I. Kant, 1929. *Critique of Pure Reason*. New York: St. Martin's Press; I. Kant, 1991. *Metaphysics of Morals*. Cambridge: Cambridge University Press.

2 There is more than animal suffering

The animal liberation movement, among other goals, seeks an end to the use of animals for food. Philosophers who write against animal exploitation agree on the goal but differ in their approaches: Deontologists argue that rearing animals for food infringes animals' inherent right to life. Utilitarians claim that discontinuing the use of animals for food will result in the maximization of utility. Virtue-oriented theorists argue that using animals for food is an unvirtuous practice. Despite their different approaches, arguments for vegetarianism or veganism have a common step. They move from the notion of suffering to the conclusion of vegetarianism or veganism. In this paper I suggest that the notion of animal suffering is not necessary in order to condemn the practices of animal farming. I propose the possibility of defending vegetarianism or veganism on the basis of arguments that do not rest on the notion of animal suffering, but rather rely on aesthetic principles, the avoidance of violence, and preservation of the environment.

The modern animal liberation movement is fueled by the work of Richard Ryder, Peter Singer, Tom Regan, James Rachels and many others.[1] These philosophers have proposed arguments in favor of vegetarianism or veganism on the basis of utilitarian, deontological, or natural rights principles, or some combination of these theories. For example, Ryder relies on the principle of "Painism," combines the utilitarianism of Singer with the rights theory of Regan. As Francione writes, "Painism, a doctrine developed by Richard Ryder, purports to combine rights- and utility-type considerations by combining Singer's emphasis upon pain with Regan's concern for the individual."[2] These arguments are well known to anyone who has even the slightest interest in animal ethics, and so it will not be necessary to spend time rehearsing them here. Furthermore, some argue for veganism on the basis of a virtue-oriented ethic. I argue that the correct way to approach the question of our responsibility toward animals is to frame the issue in terms of whether the practice of rearing animals for food are virtuous and unvirtuous. Rather than arguing that animals have inherent rights or that they should factored in a utilitarian calculus, it is more productive to realize that raising animals for food, quite apart from rights and utility, is unvirtuous: e.g., callous, intemperate, self-indulgent, and base.

Although each of the theories sketched above proposes its unique approach to animal ethics, all of them have a common linchpin, which is the notion of animal suffering. Utilitarianism, painism, subject-of-a-life, and virtue-based theories all share in common a move from the notion of suffering to the conclusion of vegetarianism or veganism. In what follows, I would like to suggest that there are other promising avenues to explore, besides the notion of animal suffering, that may lead to the same conclusions, i.e., that we should condemn animal farming and that we should become vegans or vegetarians. I do not intend to undermine the importance of animal suffering as a criterion for respecting animals. I believe that animals *do* suffer. However, an interesting question that this paper explores is, "If we discovered that animals do not suffer at all, would there be other arguments against animal exploitation?" I argue that there are other important resources that can be used to condemn the practice of intensive animal agriculture. These resources are what I term aesthetic- and gustatory-based. I shall discuss them in order.

Aesthetic-based argument

The aesthetic-based argument is twofold: the first aspect has to do with the unaesthetic nature of intensive animal agriculture. The second aspect pertains the unaesthetic nature of the violence involved in intensive animal agriculture. The first part of the argument can be expressed as follows:

1   We ought to eliminate those practices that produce unnecessarily repugnant sights, sounds, and odors.
2   Intensive animal farming causes unnecessarily repugnant sights, sounds, and odors.
3   Consequently, we ought to eliminate intensive animal farming.

The animal ethics literature has not completely ignored aesthetic concerns as promising ways to counter intensive animal agriculture. However, aesthetic principles are often overlooked. There are important philosophical arguments that can support an aesthetic-based argument against animal agriculture. Consider for example Mary Midgley's notion of the "Yuk Factor." Midgley writes, "it is especially unfortunate that people often now have the impression that while feeling is against them, reason quite simply favors the new developments."[3] Here Midgley is discussing the emotional reaction of people to certain forms of biotechnology. She points out that in moral reflection, it is usual to favor reason and discount emotion. But it is a mistake to dismiss emotion forthright. The emotion-vs.-reason dichotomy is not exactly accurate. After all, emotions often are pre-rational, but not irrational, reasons. When a practice is universally, or nearly so, repugnant, we should not dismiss our emotional reaction and treat it as morally unimportant. The "yuk factor" is the not-yet-articulated moral reaction toward certain practices that must be seriously considered, because it requires time to rise to the rational level and be articulated. As Midgley points out,

Feelings always incorporate thoughts—often ones that are not yet fully articulated—and reasons are always found in response to particular sorts of feelings. On both sides, we need to look for the hidden partners. We have to articulate the thoughts that underlie emotional objections and also note the emotional element in contentions that may claim to be purely rational. The best way to do this is often to start by taking the intrinsic objections more seriously. If we look below the surface of what seems to be mere feeling we may find thoughts that show how the two aspects are connected.[4]

Regarding the treatment of animals, the nearly uniform aversion that people experience to blood, bodily fluid, bad odor, and other "yucky" aspects characteristic of animal rearing and meat production, is not merely subjective. This aversion denotes important moral implications. Even many meat eaters show repugnance toward the processes involved in meat production, from the squalid rearing aspect of animals to their slaughtering. Not surprisingly, slaughterhouses are hidden from plain sight. And also, not surprisingly, meat eaters typically avoid acknowledging from where their meat comes.[5] Meat eaters consume cooked and seasoned, rather than raw, animal flesh. As Kuhen notes, people like their vegetables to look like vegetables, but do not like their meat to look like animals.[6] The simple explanation is that the process of rearing animals, slaughterhouses, and the final products, the animal parts that people consume, are inherently repugnant.

A. G. Holdier proposes a version of an aesthetic-based objection to meat. Holdier presents an interesting argument inspired by the writings of Henry Stephens Salt.[7] Writing about the immorality of slaughterhouses in the late 1800s, Salt is a pioneer of animal rights and vegetarianism. As Holdier points out, Salt's arguments are not formal, though represent the beginning of a promising aesthetic-based argument: "Slaughterhouses are disgusting, therefore they should not be promoted."[8] Holdier seeks to build an argument upon Salt's aesthetic considerations regarding slaughterhouses. Two important steps in Holdier's argument are, first, the fact that the conditions in factory farms and slaughterhouses are objectively dreadful. It is not necessary to present an argument to show that animal farming is disgusting. Nowadays, the horror of slaughterhouses and animal suffering in general are very well documented. There exist numerous YouTube video exposés as well as articles and books describing and show the horrific conditions of animal farms and slaughterhouses. The second step is to suggest a moral dimension to aesthetic judgments to shows that the repugnant often signals that something is immoral or bad for us.[9] The connection with morality and aesthetics is, obviously, not unusual. Holdier proposes such a connection in the light of recent work in the psychology of disgust, which suggests that our reaction to the ugly is a warning that something is dangerous or wrong for us.[10] Consequently, it is plausible to assume that our repugnance toward the conditions of farm animals and the ghastly processes involved in the production of meat stems from an internal cognitive mechanism that recognizes the wrongness of such practices.

Furthermore, consider that there are certain mechanisms used to subvert our opposition to animal exploitation.[11] For example, early on in our lives we have to be conditioned to regard animals as food. As children, animal food is presented in forms that do not remotely resemble animals, such as mush or nuggets or meals labeled "happy." As Josephine Donovan correctly notes, "Children have to be educated out of the early sympathy they feel for animals."[12] Children are kept uninformed of the process required to turn animals into their "happy meals." Children's books often depict animals as happy friends, rather than showing them amassed in cages inside factory farms. Virtually all children are not taught that burgers and steaks are body parts of the same cute and loving animals about which children see in their books. There is a clear mechanism that disconnects children's understanding of the lives of animals. As Carol J. Adams notes,

> We live in a culture that has institutionalized the oppression of animals on at least two levels: in formal structures such as slaughterhouses, meat markets, zoos, laboratories, and circuses, and through our language. That we refer to meat eating rather than to corpse eating is a central example of how our language transmits the dominant culture's approval of this activity.[13]

An aesthetic-based argument, therefore, considers the aesthetic value loss that the practices required to produce meat causes. The point of the argument is that a life that contains a lesser amount of unpleasant sights sounds and odors is more conducive to flourishing than a life filled with unaesthetic sights and violent events. The world has many terrible features, such as natural diseases, crime, pollution, discrimination, and more. Fortunately, the world still has many positive and beautiful characteristics. For example, when we contemplate nature, unless one is completely insensitive about it or unmoved by it, we find objective beauty. Animals and insects represent one of the many beautiful aspects of nature. They are unique and render nature beautiful just like the elements that constitute a great painting. Imagine how sad the world would be without the colors and sounds of animals. It would be comparable to the painting Mona Lisa without the Mona Lisa. Granted, the meat industry does not intend to wipe out animals from the world. However, my point is that the meat industry affects negatively the beauty of nature damaging it and adding violent and unpleasant features in the world. Consider, for example, how cutting down trees to build a gas station, oil spills in the ocean, pollution, and more, make life less enjoyable. Moreover, consider the terrible loss when some animals become extinct. Even insect extinction would have a negative impact on plants and consequently grains, vegetables, and fruit on which humans and animals rely.

Raising animals for food undermines the natural beauty of the world by domesticating and bringing into existence millions of animals for the purpose of slaughtering them and using their bodies for food. Note that this line of argument is not committing the naturalistic fallacy. The argument is not that what is natural is necessarily good. Rather, *certain* aspects of the world and certain human practices make life less enjoyable and less aesthetically pleasing. Apart from the fact that animals suffer, it is necessary to realize that the practice of raising animals for food contributes to the ugliness of the world. What is ugly about raising animals for food is the fact that millions of animals have to be brought into existence and slaughtered; it is known that this is the cause of global warming, deforestation, pollution, and other problems that affect the ecosystem.[14] Also, it is hardly necessary to give a sophisticated argument to show that slaughterhouses are aesthetically unpleasant places—inside and outside—not just for the animals but also for humans who see or know about slaughterhouses and for the humans who work in them. What happens inside slaughterhouses cannot be said to add to the beauty of the world. There is nothing beautiful about a world of trucks loaded with animals that are forced into squalid slaughterhouses where the animals enter whole and exit in pieces. There's nothing beautiful about the foul odor that slaughterhouses emanate or the waste that is produced by the slaughtering and preparation of meat. Moreover, we have to consider that regardless of slaughterhouses killing animals is not an aesthetically pleasant practice to perform or watch—even for meat eaters.

The beauty of animals to which I am referring is evident by the way that many people have companion animals. They find them lovable and cute and for many these animals fill an important gap in their hearts. As some have pointed out it seems arbitrary that some animals are treated as companions and others as food.[15] In fact, besides cats and dogs, also pigs, cows, chickens, and others are lovable animals. Suppose that the overwhelming majority of scientific studies conclusively showed that animals do not feel pain. Descartes, for example, was convinced that animals were like nature's robots. He argued that animals behave and look as if they were sentient, but, in reality, they are merely biomechanical machines. Assume that Descartes was right. Would people no longer become attached to companion animals? Would people abandon their companion animals just because these animals lack the capacity to feel pain? Would animals not matter any longer? I think it is clear that the answer is no. We contemplate the beauty of sunsets, the northern lights, the forests, and the oceans. Their beauty enriches us despite not being sentient things. By the same token, even under the assumption that animals do not feel pain, it seems to me that it would not change the fact that they are beautiful and their beauty, company, and presence in the world gives us an objective reason to not destroy them or raise them for food. It seems to me that regardless of whether animals can suffer, the practices involved in rearing and slaughtering animals are esthetically unpleasant features of the world, which is exactly what everyone—meat eaters and vegetarians—wants to avoid or eliminate. Thus, the ugliness of slaughterhouses, animal overpopulation, and the damage to the environment caused by animal agriculture could constitute grounds for opposing raising animals for food, irrespective of animals' capacity to suffer.

A possible objection is the following: It is difficult to separate out the aesthetic concern regarding meat production from all the others. Almost any process in an industrialized system is going to be pretty ugly. But if a given process didn't create a number of negative externalities, and people liked the product, then the ugliness would be a very weak reason to abandon that product for some alternative—likely not strong enough to condemn the practice of intensive animal agriculture, as I suggest. It is true that many processes may be considered ugly. I want to offer two observations: First, there is a significant difference between the ugliness of, say, a giant metalworking factory or cleaning septic tanks and what goes on in slaughterhouses. As I have discussed above, slaughterhouses provoke disgust virtually universally, while metalworking factories do not. There are important evolutionary and psychological reasons that ground our disgust toward blood, bodily fluids, and other details involved in animal farming. However, I am not here to argue that metalworking factories or cleaning septic tanks are not ugly. Arguably they are, but some practices are more necessary and less ugly than others. And certain processes or aspects of life, such as intensive animal agriculture, can be controlled or eliminated; other processes, such as making cars or cleaning septic tanks, are not easily avoidable because are necessary aspects of our lives. Second, my aesthetic-based argument comes with a qualification, that we should condemn an ugly activity or practice whenever viable alternatives exist. Since vegetarianism and veganism are viable alternatives to intensive animal agriculture, we should condemn intensive animal agriculture.

Furthermore, intensive animal agriculture does create a number of negative externalities, that is, a negative impact on the environment. I am referring to the environmental degradation caused by the various processes employed by animal agriculture. Every year, over 48 billion animals are slaughtered in the United States alone. This large number means, among many other things, more water usage, more animal waste, more pollution, more deforestation, more energy consumption, and the exacerbation of the already serious issue of global warming.[16]

Consider how much unnecessary cruelty animal-based diets generate. First of all, human beings do not need to consume animal products to survive.[17] Thus, in many affluent countries consumption of animal products is not a necessity. Factory farms and slaughterhouses are very unaesthetic. Not surprisingly, slaughterhouses are hidden away in rural areas away from eyes and ears and noses. Most people would readily admit that rearing and killing animals produce a great deal of displeasing sights, sounds, and odors. The prospect of thousands of animals amassed in confined spaces is a squalid sight that no person with a sense of aesthetic would bear to see. The animals are forced to live against their nature. They contract diseases and die. Consider also the squalor of slaughterhouses, covered in blood and entrails. Consider the workers covered in blood killing thousands of animals on a daily basis. Consider the instruments required:

Stunning Pen (also referred to as the knocking pen): A narrow enclosure in which the animal is rendered unconscious.

Skinning Cradle: A metal or plastic rest for skinning and eviscerating animals.

Collecting Troughs: These are containers for receiving blood or collecting gut material.

Sticking Knife: A knife with a six-inch blade and a v-shaped end used in severing the blood vessels of the neck to bleed the animal.

Skinning Knife: A knife specifically designed to remove the animal's skin.

Meat Saw: A handsaw used to saw through bone of animals.

Meat Cleaver: A heavy axe used for separating heavy structures, such as the head from the neck.

Meat Tree/Hooks: Metal devices with curved ends for holding parts of the slaughtered meat.

Judging by the facilities, the equipment, and the practices of animal farming, it is clear that they involve violence that undermines beauty and introduces ugliness in the world. Conversely, the human diet does not require cruel practices or unaesthetic practices. Compare bone saws cutting through animal bones, cutting an animal's throat with a sticking knife, skinning, and the stench emanated from such a process with picking ripe fruit from a tree, slicing a melon, or uprooting a head of lettuce.

In short, raising animals for food is revealing unsustainable as a practice; it contributes to the degradation of the environment.[18, 19] The point here is that animal suffering is not necessary to denounce the practice of raising animals for food. It seems clear that moral integrity demands that we avoid raising and killing animals for food to avoid an environmental disaster. This last point, however, does not apply globally, but rather to affluent societies that use intensive animal farming.

Integrity-based argument

The first part of the aesthetic argument explained in the previous section shows that there is *pro tanto* reason to condemn intensive animal agriculture because the practices of raising and slaughtering animals are unnecessarily unaesthetic. The second aspect of the aesthetic argument relies on the notion of *non-violence*. Arguably one of the most daunting aspects of society is the constant prevalence of aggression and violence. The argument is the following:

1 Non-violence is a virtue in and of itself.
2 Unnecessary violence ought to be avoided or eliminated.
3 Intensive animal agriculture produces unnecessary—yet avoidable—violence.
4 Therefore, intensive animal agriculture ought to be avoided or eliminated.

Non-violence as a moral principle has been used as a peaceful way to attain political and social change by the likes of Jesus, Socrates, St. Francis of Assisi, Mahatma Gandhi, Martin Luther King Jr., Lydia Maria Child, Violet Oakley, and others.[20] What I argue here is simply that violence toward animals is recognized by many, and for obvious reasons, to be wrong because it hurts animals. However, apart from causing harm to animals, I argue that violence is inherently wrong. A similar argument was used by St. Pius V who issued a papal bull titled "De Salute Gregis Dominici" to ban bullfighting. Pius V was concerned about the danger of fighting animals; moreover, he was concerned about human souls. He understood that violence undermines human dignity. The violence involved in the operations of the meat industry, should be condemned for the same reason.

Timothy Pachirat, was still a Ph.D. student at Yale University when he gained employment in a slaughterhouse in Omaha for the purpose of writing his dissertation. At the slaughterhouse, his jobs were liver hanger, cattle driver, and quality control. During his employment in a slaughterhouse, he wrote what would later become the book *Every Twelve Seconds*.[21] Chapter 3 is the most difficult reading of the book because it describes in details the violence he experienced while working at the slaughterhouse. Pachirat describes the fear of the cattle as they approach the time of their death. After they are killed, workers cut the animals into pieces and remove the animals' organs. Pachirat describes a slaughterhouse where thousands of animals are killed every day; here the stench of animal cadavers, the blood, the entrails, the fear of the animals are magnified compared with, say, a small farm where a farmer kills an animal to feed his family. But can killing an animal and preparing her flesh for consumption be done gently? Aside from the number of animals killed, in what sense does killing one or many differ? Some said that if slaughterhouses had glass walls, everyone would become vegetarian. One of the main reasons, in my view, is the useless violence of killing animals for food.

Jan Deckers also describes how animals like chickens, pigs, and cows are confined in overcrowded and filthy quarters.[22] Animals are overfed and develop metabolic diseases. Also, they develop infections due to the unsanitary conditions in which they live. Every decision is made in consideration of profit rather than the welfare of the animals. One example worth mentioning is the chicken farm's treatment of male chicks. Since males do not lay eggs, chicken farms kill them immediately after hatching. They are killed by means of suffocation in large plastic bags, gassed to death, or ground alive. Chickens are then hung from their feet and their heads dunked in electrified water to cause cardiac arrest. They are then decapitated and dunk in boiling water. Those that are still alive drown in the water. Virtually all people (except the deranged) would be appalled at such practices. I will spare the reader from the details of slaughtering cows and pigs. Somehow the fact that they are larger animals, with more flesh, entrails, and a lot more blood than chickens, makes their slaughtering even more gruesome. Some people, however, argue that it is possible to kill and eat animals when it is done in a "humane" way. Such a comment never ceases to baffle me. I wonder how it is actually possible to be humane and at the same time decapitate, gut, bleed out, and cut up any animal (human or non-human).

Some readers will immediately retort that killing is not always wrong. Perhaps that is the case, but my argument is that useless violence is always wrong, and we are better off without it. Killing and then slaughtering an animal for food inherently involves violence. The details of killing for food a placid creature such as a cow or a gregarious animal such as a pig will bring most sensitive people to tears and make them sick to their stomach. This is not a fallacious appeal to pity. My argument is not that killing cute and innocent creatures is immoral because they are cute and innocent. My point is that killing and slaughtering involves violence, and violence is not moral or desirable. One may point out that there are exceptions in extreme circumstances, say, when it is necessary to save lives. I am not arguing that because violence is wrong we should never use violent means in the face of extreme danger. The point that I want to make is that most people would be consternated by the killing, death, and slaughtering of animals regardless of whether animals can feel pain. It is not (necessarily) our understanding that animals have the capacity of suffering that makes us cringe at the idea of killing and preparing an animal for consumption. It is the overt violence of using instruments such as knives, hammers, nail guns, and more to cut, pierce, smash, tear. It is the instruments used to kill, the stench of death, the horrid sight of blood and entrails that follows the death of an animal.

Granted, it is not the same for everyone. These aspects may not affect some people in the slightest. But most people would refuse to kill an animal for food because of the violent acts necessary to kill and cut up the flesh of that animal. Compare slaughtering with peeling an orange or taking some lettuce from the garden. Eating vegetables does not involve violence. After all, it seems to me that slaughterhouses do the dirty job that people would otherwise hate to do—and that's why their walls are not made of glass. Meat eaters, in a sense, delegate this violence to slaughterhouses. Now regardless of whether animals can suffer like humans or in a similar way or not at all, the violence involved that I describe is an objective reason for opposing to raising and killing animals for food. Most people would feel uncomfortable performing such violent acts as killing an animal and cutting up, slicing, deboning the flesh of that animal. This is the first aspect of integrity to which I am referring.

Another way to understand the intrinsic value of nonviolence is to point out that violence generates more violence. Thomas Aquinas and Immanuel Kant both discuss this aspect. Thomas Aquinas argued that despite their capacities that are similar to those of men, it is not a sin to kill animals because God created them for the benefit of humans. However, in *Summa Contra Gentiles* (1480), Aquinas states that animals exist for our benefit. However, this fact does not warrant animal cruelty. In fact, cruelty to animals should be avoided. He writes, "If a man practice a pitiful affection for animals, he is all the more disposed to take pity on his fellow-men."[23] Immanuel Kant echoes Aquinas. In

Lecture on Ethics, Kant states, "he who is cruel to animals becomes hard also in his dealings with men."[24] The point that Aquinas and Kant make seems to me to be plausible. As Sarah Watts writes, "Jeffrey Dahmer. Ted Bundy. David Berkowitz. Aside from killing dozens of innocent people (combined), these men—and a significant percentage of other serial killers—have something else in common: Years before turning their rage on human beings, they practiced on animals."[25] Consequently, even in this case, it is possible to condemn intensive animal agriculture independently of the notion of suffering. As Aquinas and Kant point out, we should treat animals with respect because being cruel to animals makes humans become cruel to each other. Following this line of reasoning, combined with what I said earlier about the violence involved in killing and preparing animals for food, it seems to me that yet another reason to oppose the practice of using animals for food is that it raises our threshold for cruelty.

Gustatory-based argument

Many philosophers arguing in favor of vegetarianism point out that food flavor should be regarded as subordinate to the life and well-being of animals. Meat eaters (and many vegetarians, as well) remind us all the time how good meat tastes and, consequently, how hard it is to give it up. However, many recognize that the taste of meat and experience of eating it are, morally speaking, less important than animal suffering. Thus, they conclude that since taste is trivial in comparison with animal suffering, we ought to give up eating meat. Others maintain that taste is not trivial at all, but rather an important value to a good life that may justify rearing animals for food. Eating well, they argue, is a significant value to humans, and eating meat is a significant part of eating well. Consequently, not eating meat is a significant value-loss.

I do not dispute that eating well and the role of flavor are important values to humans. However, I have three objections: First, plant-based food tastes good—even superior to meat according to many meat eaters who have become vegans. I think it is important to consider that many vegans and vegetarians, who formerly consumed meat, and thus know both worlds, prefer the taste of plant-based food.

Second, the notion that not eating meat is a significant value-loss to humans is an exaggeration. Perhaps, the "not eating meat is a value-loss to humans" argument would have some force under these conditions: if taste of meat were somehow *essential* to humans in the sense that without it humans would get ill or life would be utterly unbearable; or if meat were the only food that tastes good. Clearly neither is the case. Humans can easily adjust their taste, and plant food tastes just as good or even better than meat. Furthermore, considering the aesthetic-based argument, and given the environmental argument, it is sensible to adjust one's taste to plant-based food, which avoids and prevents health problems and environmental degradation.

Third, meat is not inherently good. Meat dishes become delicious under the chef's skillful hands. I anticipate resistance here. It may be objected that this applies to vegetarian food as well. Also, it may be pointed out that flavor is subjective. I argue that it can be shown that vegetarian food is inherently flavorful while meat is not. Fruit and greens and even grains don't require special preparation or seasoning. Mangos, bananas, watermelons, spinach, peppers, and more, are flavorful in their raw state or just by minimally cooking them. There are vegetables that when raw have little taste or cannot be eaten. For example, broccoli and eggplants are not ideal eaten raw. However, they are not repulsive and with very simple cooking methods, such as steaming, they become flavorful. On the other hand, meat is foul when raw and requires certain steps necessary to render it edible. In fact, unlike vegetables, meat is the flesh of once living animals now cut and shaped in ways that do not resemble animals. Also, meat requires maturation. Maturation means that the flesh of a slaughtered animal is aged for at least a few days, sometimes up to several weeks. This process is necessary to tenderize the tough muscle fibers. Furthermore, meat is never consumed as is. With very few exceptions, people would never kill, say, a cow or pig or chicken, carve out the flesh and consume it on the spot. Meat is always aged, marinated, seasoned, and cooked. For example, consider a popular dish typical of the Italian region of Piedmont called *brasato*. This dish is cattle flesh braised in red wine and spices for hours. The point is to render the meat tender and allow it to acquire the taste of the wine and the spices because it is tough and unpleasant tasting. In other words, taste is conferred upon the meat by the wine and the spices and through hours of cooking. Another dish typical of the Italian region of Bologna is a sauce with ground beef known as *ragú*, not to be confused with the Italian-style American brand Ragú. This sauce is prepared by sautéing the ground meat in oil and adding to it, again, wine. The reason is to fix the foul taste of the animal flesh and make it taste interesting with the use of wine and spices and transform it into a dish. After all, meat is, like it or not, decomposing flesh. It is true that vegetables and fruits decompose, too, but, in that case, we avoid consumption and discard them. All meat is prepared with some kinds of powerful spices, oils, or wine, and is cooked to modify its naturally foul flavor.

With regard to taste, consider the study of Anderson and Feldman Barrett, who tested whether people's beliefs of how animals are raised can influence their experience of eating them.[26] Samples of meat were accompanied by respective descriptions of their origins and treatment of the animals on factory farms. Some samples were said to be the product of factory farm, while others were labeled as "humane". In reality, all the meat samples were identical. Interestingly, the participants of this study experienced the samples differently: meat described "factory farmed" was perceived as looking, smelling, and tasting as less pleasant than "humane" meat. The difference was even to the degree that factory farmed meat was said to taste more salty and greasy than "humane" meat. Furthermore, the participants who were told that they were eating factory-farmed meat consumed less of the sample. According to the authors of this study, "These findings demonstrate that the experience of eating is not determined solely by physical properties of stimuli—beliefs also shape experience."[27] My point is that meat is not consumed because it is good in itself. It is rendered good by masking or modifying its original flavor with the use of potent spices or liquids used to marinate, season, and cook the meat. Furthermore, meat is enjoyed because it is eaten socially, during holidays for example. The taste is not consistent and not exactly pleasant, as the study just mentioned explains. Thus, my argument is that,

1 Meat is not inherently flavorful, but rather unappetizing.
2 Whatever is inherently unappetizing should not be consumed.
3 Therefore, meat should not be consumed.

I think that there is good evidence to show that the first premise is true. Also, most meat eaters would, in good conscience, concede this much. The second premise needs to be unpacked, especially in light of an obvious objection: it might be objected that it is irrelevant whether a given product is good in itself. What matters is how it contributes to a dish. And most people seem to think that animal products make dishes taste better than they would without them. Firstly, premise 2 is not meant to apply universally. It applies to affluent societies where people have easy access to an abundance of nutritious plant-based food. I do not claim that it applies to a circumstance where one, say, is stranded on a desert island, compelled to eat a food that provides sustenance but is not tasty.

Second, the reason we should avoid food inherently unappetizing in the case of meat is to avoid self-deception. That is, since meat is not inherently tasty, as I discussed above certain steps must be taken to mask its unpleasant taste, appearance, smell, and texture by curing, seasoning, and cooking it. Taking such steps constitutes an act of self-deception. Since one should avoid deceiving oneself, it follows that one should avoid eating meat. It is not the taste of meat itself that meat eaters like, but rather the taste of the seasoning, spices, and flavors created by the cooking process. Thus, realizing these facts may constitute a reason for one to avoid eating meat. I can see that this argument may not convince many, especially meat eaters. However, the main point is that it need not convince all. As Allen W. Wood aptly puts it,

> Some philosophers seem to think that each proposition in a theory must be argued … using arguments that are supposed to persuade anyone at all, even someone with no sympathy whatever for the project in which the theory is engaged. That is a standard that no philosophical theory could ever meet.[28]

In fact, I do not pretend to persuade anyone at all, but rather those sympathetic with the project of condemning factory farming and supporting vegetarianism or veganism on the basis of principles that do not rely on the notion of animal suffering. With this consideration in mind, the above argument accomplishes such a goal.

Conclusion

In this chapter, I argued that a common trend in animal ethics is the move from the notion of animal suffering to veganism or vegetarianism. The argument is roughly the following: those animals used for food are said to be sentient, i.e., they have the capacity to suffer. Causing unnecessary suffering is wrong. For those of us who are fortunate to live in places where plant-based food abounds, killing animals for food is unnecessary. Consequently, killing animals for food is wrong. Obviously, philosophers present this argument with several intermediate steps. At any rate, the point of this paper is that animal suffering is not all that matters. If it were discovered that animals do not feel any more pain that rocks, would there be other moral considerations regarding using animals for food? I maintained that there are other arguments that can be used to condemn the practice of intensive animal agriculture and support vegetarianism or veganism. These arguments are aesthetic-, integrity-, and gustatory-based.

First, the practices required to rear animals and transforming them into food are aesthetically repugnant. Our natural aversion to practices that are aesthetically distasteful, such as slaughtering, carries moral implications sufficient to condemn animal agriculture. This argument has two ramifications: one, we want to avoid unnecessary ugliness in the world, and animal agriculture brings about a great deal of unnecessary ugliness. Our natural aversion to blood, flesh, and other repugnant aspects involved in animal food production signal that something is bad for us. Its deleterious effects on human health further corroborate the badness of animal food consumption. Second, animal agriculture is inherently, and unnecessarily, violent—and it instills in us violence. Since we should value non-violence as a virtue, it follows that we should avoid those practices that are inherently, and unnecessarily, violent, such as meat production in the presence of an abundance of plant-based food. Furthermore, a valid reason to embrace veganism, aside from animal suffering, is to realize that meat is not inherently tasty. Rather, meat is rendered palatable by masking its naturally foul taste with various spices, condiments, and cooking techniques. Since eating such foods is a form or self-"deception, I argue that we should avoid them.

Notes

1 The respective relevant works I refer to are the following: Tom Regan, *The Case for Animal Rights*. Berkeley, CA: University of California Press, 1983; Richard Ryder, *Speciesism: The Ethics of Vivisection*. Scottish Society for the Prevention of Vivisection, 1974; Richard Ryder, *Animal Revolution: Changing Attitudes towards Speciesism*.

Cambridge: Cambridge University Press, 1989; Richard Ryder, *Painism. A Modern Morality*. Open Gate Press, 2003; Richard Ryder, "Painism." In M. Bekoff (ed.), *Encyclopedia of Animal Rights and Animal Welfare* (pp. 402–403). Santa Barbara, Cal. [etc.]: Greenwood Press, 2010; Peter Singer, *Animal Liberation: A New Ethics for our Treatment of Animals*, New York Avon Books, 1975; James Rachels, *Created from Animals: The Moral Implications of Darwinism*. Oxford University Press, 1990.
2 Gary Francione, *Rain without Thunder: Ideology of the Animal Rights Movement*. Philadelphia: Temple University Press, 1996: 224.
3 Mary Midgley, "Biotechnology and monstrosity: Why we should pay attention to the 'yuk factor'," *Hastings Center Report* 30(5), 2000: 7.
4 Midgley, 2000: 8.
5 J. R. Kunst and S. M. Hohle, "Meat eaters by dissociation: How we present, prepare and talk about meat increases willingness to eat meat by reducing empathy and disgust." *Appetite*, 105, (2016): 758–774.
6 G. Kuehn, "Dining on Fido: Death identity, and the aesthetic dilemma of eating animals." In E. McKenna and A. Light (eds), *Animal Pragmatism: Rethinking Human-Nonhuman Relationships*, Bloomington: Indiana University Press, 2004: 228–247.
7 A. G. Holdier, "The pig's squeak: Towards a renewed aesthetic argument for veganism." *Journal of Agricultural and Environmental Ethics*, 29(4), 2016: 631–642.
8 Ibid., 631.
9 Ibid., 633.
10 H. A Chapman and A. K. Anderson, "Trait physical disgust is related to moral judgments outside of the purity domain." *Emotion*, 14(2), 2014: 341–348.
11 Brian Luke, "Justice, caring and animal liberation." *Between the Species*, 8(2), 1992.
12 Josephine Donovan, "Feminism and the treatment of animals: From care to dialogue." *Signs* 31(2), 2006: 323.
13 J. Carol Adams, *The Sexual Politics of Meat: A Feminist-Vegetarian Critical Theory*. Bloomsbury Academic, 2015: 47.
14 Bibi Van Der Zee, "What is the true cost of eating meat?" www.theguardian.com/news/2018/may/07/true-cost-of-eating-meat-environment-health-animal-welfare
15 See Melanie Joy, *Why We Love Dogs, Eat Pigs, and Wear Cows: An Introduction to Carnism*. Conari Press, 2011; and Melanie Joy, "Dis-ease of the heart: The psychology of eating animals." *Forks Over Knives*, May 23, 2012. www.forksoverknives.com/dis-ease-of-the-heart-the-psychology-of-eating-animals/#gs.=G3q7lw
16 D. Pimentel and M. Pimentel, "Sustainability of meat-based and plant-based diets and the environment." *The American Journal of Clinical Nutrition*, 78(3), 2003: 660S–663S. doi:10.1093/ajcn/78.3.660S
17 For an in-depth discussion of why animal-based food is not necessary to good health, see Stephen Patrick Kieran Walsh, "Why foods derived from animals are not necessary for human health." In Linzey, C. (ed.) *Ethical Vegetarianism and Veganism*. Oxon: Routledge, 2018: 19–33. doi:10.4324/9780429490743-2
18 Felicity Carus, "UN urges global move to meat and dairy-free diet." *The Guardian*, 2010. www.theguardian.com/environment/2010/jun/02/un-report-meat-free-diet
19 B. Walsh, "The triple whopper environmental impact of global meat production." *Time*, 2013. http://science.time.com/2013/12/16/the-triple-whopper-environmental-impact-of-global-meat-production/
20 Women Champion Peace and Justice through Nonviolence. (n.d.). www.library.georgetown.edu/exhibition/women-champion-peace-justice-through-nonviolence
21 Timothy Pachirat, *Every Twelve Seconds: Industrialized Slaughter and the Politics of Sight*. Yale University Press, 2011.
22 Jan Deckers, *Animal (De)liberation: Should the Consumption of Animal Products Be Banned*. London: Ubiquity Press, 2016.
23 Aquinas, 2016, I-II 99: 1.
24 Immanuel Kant, *Lecture on Ethics*, 1963: 24.

25 Watts, S. (2018). "First they tortured animals, then they turned to humans." A&E. www.aetv.com/real-crime/first-they-tortured-animals-then-they-turned-to-humans
26 E. Anderson and L. Barrett, "Affective beliefs influence the experience of eating meat." *PLoS ONE*, 11(8), 2016: e0160424. doi:10.1371/journal.pone.0160424
27 Anderson and Barrett, 2016: 16.
28 Allen W. Wood, *Kantian Ethics*. Cambridge University Press, 2008: 1.

3 In vitro meat

Is cultured meat a solution to animal suffering and environmental degradation?

The answer that I will give to the question posed in the title is, "No, it isn't." This is my position regarding in vitro meat[1] and in this chapter I will explain why. Latest technology enables scientists to grow replacement body parts from the cells of a patient.[2] For a number of years now, scientists have been studying the possibility of growing meat in laboratories, a project commonly known to the public as the (in)famous in vitro meat or lab-grown meat. While samples of such meat exist, it is not yet commercially viable. It will probably take many years to see lab-grown meat in supermarkets. In this chapter, I want to explain why I hope that the project of synthesizing meat should not be morally supported. What is the moral status of growing meat in a Petri dish? This question includes the morality of pursuing such a scientific research and the morality of buying/supporting and consuming in vitro meat. This question is interesting especially to ethical vegans. Meat eaters eat meat and arguably many meat eaters will try and regularly consume lab-grown meat. But what about vegans? Vegans avoid eating meat, but this isn't meat. Thus, many vegans will likely consume lab-grown meat—a very strange form of veganism, to say the least.

The idea behind such a project is that if lab-grown meat can be produced in a way that avoids environmental degradation and the meat is sold at a competitive cost, the projects will be able to avoid the damages that the farm animals' sector causes to the environment. It would also spare many farm animals. It would seem, then, that the research and consumption of lab-grown meat are justified by the potential environmental benefits. Furthermore, growing meat would reduce animal suffering because no animals will be raised for food. Strange as it may sound, some vegans may support such an endeavor. In fact, some vegans may even consume lab-grown meat. After all, vegans avoid meat because to produce meat animals suffer and the environment is degraded. Vegans avoid (as much as possible) causing or contributing to animal suffering and avoid activities that degrade the environment. Since lab-grown meat has a much lower global impact on the environment, then some vegans will presumably consume lab-grown meat.

There are obvious objections that one could raise. Animals have to be used, in any case, for the production of cultured meat. Namely, whether painful or painless, animals must be reared and their cells must be taken from them in order to replicate them in a laboratory. Consequently, animal exploitation will not cease to exist completely. But perhaps completely discontinue rearing animals is not part of the project. Obviously, supporters of in vitro meat may argue that it is better to reduce animal suffering and the negative impact on the environment than to continue to exploit animals. However, we must keep in mind that leather, fur, eggs, milk, and other animal byproducts would still require raising animals. Another problem is that eating artificial meat is not risk-free. How could researchers possibly know about long-term effects of a laboratory product? If it's meat, then it's meat. That is, whether from the lab or from the farm, meat is unhealthful to humans.[3]

Furthermore, it is unlikely that all people in the world would eventually switch to eating in vitro meat. In my view, turning people to in vitro meat is more difficult than trying to make people switch to a vegan diet. Granted, it is possible that in the future lab-grown meat will taste exactly like the real deal and people will overcome the squeamishness of artificially grown meat. At any rate, it is reasonable to believe that not all meat eaters will adopt cultured meat as a staple. Consequently, I believe that lab-grown meat is not an alternative competitor to raising animals for food, but rather just another option, and thus that traditional meat production will not change to the degree necessary to save the planet from the negative environmental consequences we are now facing.

The most important aspect of this issue is that the endeavor of growing meat, the research, the resources used, and the effort made, the ingenuity of the minds being employed on such a project, are all misplaced. It is true that it is necessary to keep in mind sustainability and to try to reduce suffering in the world. But the problem is that supporters of in vitro meat overlook an important question: What if, in trying to create artificial meat, we are doing something callous, base, or unvirtuous? I want to suggest that if reducing animal suffering and avoiding the negative environmental impact caused by animal agriculture is what we try to achieve, it seems more appropriate, more practical, and morally sound to move toward an animal-free diet altogether rather than perpetuating production of meat. Still, if lab-meat one day becomes readily available, affordable, and it tastes exactly like real fresh meat, what would be the moral problem at that point? What would vegans have to say then? As I said above, it seems plausible that those vegans, for example, who avoid meat for ethical reasons, but like the taste of meat, would accept, support, and even consume lab-grown meat.

Veganism does not have to come to such an absurdity. But I also want to point out that humanity does not have to resort to such an absurd stratagem as growing meat synthetically to save the environment. This absurdity, namely, vegans who support, accept, and might even consume lab-grown meat, stems from an ethic of expediency that focuses on the best-outcome approach. Also, it stems from a lack of temperance toward food. One way to overcome moral absurdities and to see through this is to adopt an attitude that, instead of asking what's in it for us or what our duties regarding this issue are, instead ask whether the project of growing meat synthetically conforms with important virtues, such as justice, compassion, and temperance, to name a few. In the course of this discussion, I hope it will emerge that investing valuable resources in producing lab meat, supporting and consuming lab meat, are motivated by lack of virtue. Plants are abundant, environmentally friendly, and nutritious. Meat is not healthful, and there is no reason to believe that its synthetic version might be. Consequently, these premises lead to the conclusion that humans should invest the same energy and resources to create a meat-free world instead.

Lab-grown meat and ethics

I find the idea of lab-grown meat to be fundamentally wrong, morally and practically speaking. And I doubt that humans are in fact clamoring for synthesized meat. A small group of individuals and corporations are trying to shove this down the throats of the public, mostly for profit or other reasons that have nothing to do with ending animal agriculture. The discussions of the validity of lab-grown meat are typically framed on utilitarian grounds. It is not important to specify which particular brand of utilitarianism, because the central idea is that the goal is the end result. Utilitarianism as a theory argues that a moral choice is the right one just if acting on it will maximize overall satisfaction of preference (or pleasure). Critics of utilitarian ethics, sometimes, in criticizing utilitarian morality commit the straw man fallacy. In reality, utilitarians are not cold people who do not care about anything but maximum utility. At any rate utilitarianism permits actions that most people would consider, if not immoral, quite bizarre. Leading utilitarian philosophers argue that issues such as infanticide, cannibalism, bestiality, and incest, to name a few, are not inherently wrong. Whether these issues are right or wrong is a matter of their capacity to generate desirable or undesirable consequences. In other words, if doing X accomplishes the overall satisfaction of preferences, then doing X is morally required. This of course is a sketch of the theory, which is more elaborate than this—but not much more. In animal ethics, utilitarianism, we must concede, has had, and continues to have, an influential and persuasive power on people. When the present issue of growing meat artificially is assessed from a utilitarian standpoint, the result is that, given the success of the project, growing, supporting, and consuming lab-grown meat is morally required because it maximizes aggregate utility.

Deontology is more complicated for many reasons. For one thing, there are various versions of deontological ethics, though the essential argument is that morality is a system of rules typically dictated by reason. Moral actions are performed out of a deep sense of duty. There are, of course, many difficulties with such an ethical approach. The details of possible difficulties is not important here, not to mention that it would be beyond the scope of this discussion and of this book. The important point here is that it is plausible to think that a deontological approach to the issue of lab-grown meat might suggest that producing in vitro meat is morally required. On a Regan-like approach—that is, a rights theory view—in vitro meat is required because it promises to discontinue animal agriculture; this would be beneficial to animals, who are subjects-of-a-life, i.e., they have beliefs, desires, memory, feelings, self-consciousness, an emotional life. And on a neo-Kantian view, lab-grown meat is morally required as a matter of our duty toward humans and animals.

Another way to approach the same issue, however, is to assess the motivations and also to assess the moral character from which the pursuit of lab-grown meat springs. I believe that when we consider this issue in terms of character, that is, rather than determining our duty or the right outcome, we realize that the motive behind lab-grown meat is actually corporate profit and consumer self-indulgence. Consider the importance of a temperate character in life. Physical appetites are quite important, but subordinate to, for example, justice. A temperate individual is one whose approach to eating is measured. Here I am not talking about being moderate in the sense of between eating steak every day or eating steak once a week the moderate choice is eating steak twice a week. Naturally, a consistent account of moderation must account for the circumstances. As we have seen, it is exactly animal agriculture that led us to the dire environmental issues that we are now facing. Furthermore, we must consider that the health of humans has been getting worse and worse; that humans can actually do better when they eliminate animal products from their diet; and that the taste of meat is not so important as to make one discount environmental issues. Consequently, it is plausible to say that moderation, with regard to animal products, means switching to fruit and vegetable diets.

Ethical veganism and lab-grown meat

The supporters of lab-grown meat claim that producing this type of meat could be the solution to the negative environmental impact of animal agriculture and it could reduce, or possibly end, animal suffering. Some of the comments by people on vegetarian forums are very telling: "I hope that lab meat someday catches on and can help end animal suffering. However, I would probably gag." writes one user.[4] Another user writes, "I'm a vegetarian because of cruelty of factory farming and because of the terrible environmental impact it causes, so yes, I would eat the meat."[5] The issue of whether lab-grown meat is moral seems to be addressed in terms of taste, concern for the environment, and animal suffering that can be caused by animal farming. Taste does not seem to be so important to justify animal exploitation, though as we will see it is a vital aspect that will determine meat eaters' acceptance of lab-made meat. However, environmental concern and animal suffering are very serious issues. About the environment, in 2013, Dr. Mark Post's laboratory in Maastricht University in the Netherlands produced the first lab-grown burger. Dr. Post argued that growing meat in labs could reduce the negative impact on the environment cause by the farm animals' sector.[6] If that were the case, from an ethical point of view, it would seem that even the staunchest ethical vegans (like myself, for example, who detest the idea of eating meat whether or not it is lab-grown) would have nothing to object about it. However, the morality of lab-grown meat is not as simple a question as it might appear. I will discuss some of the positive and negative aspects of the issue and later explore the possibility that considering lab-grown meat is a moral mistake that we might be able to see by embracing a virtue-oriented ethics.

Among some of the advantages of creating meat in a laboratory, the most evident is that if meat is grown in a lab, animals will not be brought into existence. Even if growing meat will not completely discontinue animal agriculture, it might mean a smaller number of animals that are raised and killed for food every year. But it is not only a matter of reducing animal suffering. Fewer animals brought into existence and raised for consumption means a less severe impact upon the environment. At present, animal agriculture contributes to 51% of global greenhouse gas emissions,[7] it uses a third of the earth's fresh water, 45% of the Earth's land, it causes 91% of Amazon rainforest destruction, and is a leading cause of species extinction, ocean dead zones, and habitat destruction.[8] The good news about artificially-grown meat, according to a 2011 study, is that, "cultured meat involves approximately 7–45% lower energy use ... 78–96% lower GHG emissions, 99% lower land use, and 82–96% lower water use depending on the product compared."[9] These numbers sound great, just what an environmentalist wants to hear; however they are theoretical since lab-grown meat is not yet a reality. At this juncture, the reality of cultured meat is unknown and so is its actual environmental impact. Although it seems plausible to speculate that under ideal conditions, all things considered, lab meat could be a more sustainable reality than the current practice of animal farming, it is just speculation.

Another point is the question of taste. Taste is a very important aspect of a good life. One interesting point about cultured meat[10] is that the meat grown for the (in)famous 2013's burger experiment was not exactly ideal by meat eaters' standards. The meat produced by Dr. Post in his lab was merely lean muscle fiber. It lacked the typical characteristics of meat, such as the fat, nerves, and blood, which are the very components that give meat its characteristic taste and texture. Creating something that resembles the "real deal" in both taste and consistency, the muscle fiber has to be "exercised" and supplied with artificial blood flow, oxygen, and nutrition.[11] These achievements are no small potato. And though I personally doubt this will be possible, it is not excluded that the project could be actualized in the future. If in the future scientists will be capable of producing lab-grown meat that is identical or nearly identical to real meat, then at that point consuming lab-meat rather than "real" meat would contribute to reduction in land usage and energy usage—provided that meat eaters would have no issue consuming lab-grown meat, that is.[12] Incidentally (or maybe not so incidentally), the typical argument that meat eaters proffer up is that eating meat is natural, it is part of *Homo sapiens*' nature as a hunter. But if we grow meat in a Petri dish, how "natural" is that? My point here is simply that it is hard to believe that meat eaters will abandon the notion of living and breathing farm animals. As a matter of fact, unless researchers are planning to lab-grow dairy products, living animals will still be needed. Furthermore, consider eggs, leather, fur, and all animal products, it seems unlikely that lab meat will have such positive effects if we consider the amount of animal products that people consume.

Considered from a consequentialist standpoint, the question of the morality of lab-grown meat would seem quite clear-cut. Assuming that in a near future lab-grown meat tastes the same as traditional meat, and it is affordable, that would be the end of the story. Under this assumption, animal suffering could be dramatically reduced or perhaps even eliminated while meat eaters would have their fix and be happy. I suggest that the story does not end there if we consider the project of growing meat from the standpoint of a virtue-oriented ethic. There are several factors to take into account before it can be said that lab-grown meat can reduce animal suffering. One of these issues is that to be cruelty-free, it might be suggested, it should be animal-free. The challenge at present is for scientist to find a method of self-renewing stem cells and animal-free materials to accomplish growth. Currently, researchers are still working on the possibility of an initial harvesting of animal cells that in the long run will no longer require subsequent harvesting. Furthermore, researchers are looking for a suitable plant-based material that will serve as "scaffolding" for the development of animal cells into a hunk of meat.

Very roughly explained, cells are taken from a living animal and allowed to grow in a Petri dish in a laboratory. In practical terms, the initial harvested cells are taken from animals that are raised according to specification so that their flesh can be genetically emulated in a lab. But is that the end of the process? Are animals off the hook after that (pun intended)? It seems not. Dr. Post points out that, "the most efficient way of taking the process forward would still involve slaughter [using a] limited herd of donor animals."[13] In other words, assuming that lab-grown meat becomes a reality, it would still be necessary to raise animals and use them as cell banks, so to speak, from which to harvest fresh cells that will be grown into meat. Granted, the number of animals involved might be smaller than the current number of raised and slaughtered animals and suffering would be minimal to no suffering at all. But how much smaller? It will not, it seems, discontinue raising animals altogether. It is difficult to imagine how it could ever be possible to produce meat in a lab without ever involving live animals. However, some argue that, eventually, it will be possible to establish a self-renewing stem cell line; that is, it will be possible to start the process by an initial biopsy and get it going without resorting to further harvesting.[14] But would it be the end of raising animals for food? Considering the wants of meat lovers, and the astute operations of the market, meat eaters might see that the sky is the limit when it comes to variety and taste. At that point, if growing meat becomes as easy as researchers hope it to be, why not clone any kind of animal meat, including, but not limited to, wild animals. In fact, creativity is perhaps humans' greatest attribute and thus it is not unlikely that some meat makers might consider extravagant cross breeding to offer a new product on the market. That's how capitalism works after all. This might require raising more animals to obtain the new kind of exotic meat desired. And again, not trying to open the proverbial can of worms here, what would be a moral objection to lab-grown human meat? After all, it would not be human meat because it would be a laboratory-grown product, but nevertheless could taste like human flesh and could be marketed as such. Labs would start harvesting cells from all kinds of animals (including, perhaps, human animals) or breeding exotic animals specially for cell harvesting, which would take us right back to square one and, once again, require breeding animals for food. Granted, it would involve an initial harvesting; but it seems that as the demand of variety increases, the number of initial harvesting processes will continue.

The next point, as I mentioned, is an issue that is seldom addressed in the discussion of lab-grown meat. This issue is the type of support that would hold the lab-grown muscle. At present, bovine fetal serum, which is an animal by-product, is used. The harvesting of bovine fetal serum is, obviously, not ideal from a vegan standpoint. Typically, the serum is obtained by sticking a needle into the heart of a fetal cow.

> At the time of slaughter, the cow is found to be pregnant during evisceration (removal of the internal organs in the thorax and abdomen during processing of the slaughtered cow).... The calf is removed quickly from the uterus [and] a cardiac puncture is performed by inserting a needle between the ribs directly into the heart of the unanesthetized fetus and blood is extracted. This bleeding process can take up to 35 minutes to complete while the calf remains alive. Afterwards, the fetus is processed for animal feed and extraction of specific substances like fats and proteins, among other things.[15]

One thing to consider here is, again, whether the end justifies the means. From the point of view of an ethic of expediency, bovine fetal serum might be just the most practical way to achieve synthetic meat. But when we assess the problem through the eyes of basic virtues, I think it is pretty evident that the achievement of an eventual synthetic support for synthetic meat is not worth the current animal torture required by the research. A study considers whether the cows can feel this procedure and the possibility of slow death from lack of oxygen, from placental separation; it also estimates that approximately 2 million fetuses are harvested annually for serum.[16] If the point of growing meat is reducing suffering, then fetal serum does not seem to be the way to go. For that reason, scientists are already working hard to find plant-based alternatives.[17]

Again, all the hard work and dedication to achieve cultured meat, and the work dedicated to finding a replacement to fetal serum with a suitable plant-based alternative, seems just disturbingly obstinate. I think there is something almost perverse about the motivation and tenacity of scientists in their pursuing lab-grown meat. The FDA said in a statement, "Given information we have at the time, it seems reasonable to think that cultured meat, if manufactured in accordance with appropriate safety standards and all relevant regulations, could be consumed safely."[18] The immediate reaction to this sort of comment, again, is why not eating vegetables and fruit, which bear no potential health issues? Other studies argue that since meat is literally grown in a laboratory by scientists, it could be possible to decide the amount and type of fat cells and other dietary characteristics of the final product. Furthermore, since slaughtering the number of animals that are slaughtered at present could be reduced, the threat of pathogens and contamination would be reduced or eliminated.[19] I think the obvious question is, "How would the FDA or researchers possibly know this?" I am not suggesting that we should not trust scientific data, but there is no way to know the long-term effects on human health of artificial meat.

To add one more "if" to the list, provided that scientists will eventually be able to grow meat without harvesting cells from living animals, find a plant-based growth-medium, reduce or eliminate animal suffering, and reduce the damage caused by animal farming, should we not abandon our moral reservations about lab-grown meat? One obvious issue can be put thus: who would want to eat meat that was grown in a laboratory? Surveys seem to suggest that the majority of people would be, to put it candidly, very reluctant to ingest lab-grown meat.[20] Granted, proponents of cultured meat remind us that "lab-grown" is a misleading term. That is to say, at present, for obvious reasons, research is being conducted in laboratories; but eventually, the meat would be made in factories. As Mattick and Allenby point out, "A world where meat comes mostly from factories instead of ranches and feedlots might be a world better able to deal with challenges of food security, the environment, and natural resources."[21] In fact, growing meat could provide a reliable and safe way to make sure that meat is devoid of hormones, antibiotics, and other chemicals that typically are given to animals.

However, some commentators still remain unconvinced. The CEO of SAFE makes a valid point,

> It is also possible that in order to overcome the public resistance to [lab-grown meat] governments and charities will be asked to fund PR campaigns and meet the research and development costs of [lab meat]. This could possibly lead to public revenue being spent on developing and promoting a technology and product that the majority of the public do not want and that will be of benefit to only those who can afford it. The Dutch government has already funded research into [lab-grown meat] conducted by New Harvest.[22]

But it might be objected that even so, funding research is justified because it will be for the benefit the environment and the benefit of the animals. After all, the argument might go, at this point it does not seem likely to hope for a vegan world. People do not seem to be interested in avoiding animal products. Certainly, continuing to promote meat even in the form of a laboratory product does not help dissipate the notion that flesh is human food. So rather than fighting against the current issue of animal exploitation, which seems to be an uphill battle, we ought to search for practical and viable solutions to replace meat—and the solution is lab-grown meat.[23] After all, that is exactly the supposed function of the various mock meats on the market. Nowadays stores carry all kinds of plant-based mock versions of meats and even cheese. The manufacturers as well as the consumers of those products point out that such products are environmentally safe. Thus, lab-grown meat should be viewed as mock meat—the best possible mock meat because it is exactly like meat (assuming, of course that scientist are able to create a perfect replica). An organization called Why Cultured Meat, for example, argue that lab-grown meat can do more for the benefit of the animals and the environment than ethical veganism could ever dream of. They argue that providing an alternative that not only looks and tastes like meat, like Tofurkey or vegan meatballs, but actually is meat, could be (again, assuming that lab-meat researchers are successful) the most effective and viable path to the desired "animal liberation." If actualized, production of lab-grown meat could mean the abolition of factory farms and hence the avoidance of suffering for millions of animals. Furthermore, if everything goes according to plans, production and consumption of lab-grown meat could mean a healthier environment.

At that point, what would be, if any, the objections of vegans to cultured meat? In the best-case scenario, assume that researchers in the future will achieve a lab-meat that is a perfect replica of real meat in taste and everything else, and, as a result, animals and the environment will benefit from it. Shouldn't vegans and animal rights activists be happy? After all, those who avoid animal-based products for ethical reasons argue that it is immoral to eat meat because it causes animal suffering and affects the environment in a negative way. I think the answers to these and similar questions are predicated upon the kind of idea one assumes about morality. In other words, it depends on the moral outlook from which the issue of cultured meat is considered. It seems to me that the question of the morality of producing and consuming cultured meat is typically dealt with by a consideration of our duty or consideration of rights. I want to suggest that in many cases trying to figure out our duty lead us nowhere. In the present issue, what is our duty? Ought we not to be saving the world from environmental degradation? Or ought we not to avoid injustice and gratuitous suffering? The answer seems obviously "yes." And if lab-grown meat is capable of doing that, then we have a duty to support lab-grown meat; consequently, according to duty ethics, that's the end of the story. But I think that a virtue-oriented approach is the correct framework to make sense of this issue because an ethics of virtue enables us to see aspects of moral issues that are simply discounted by other moral approaches.

The plight of animals

It is no longer a mystery that factory-farmed animals suffer. But why do they suffer? They suffer because they are deprived of what makes them flourish, sunlight, freedom, their natural food, their families, and more. Factory farms are squalid places where animals are often crammed into cages so small that prevent them from even moving around. In addition to the lack of adequate space, animals live and defecate in the same quarters, they are given all kinds of supplements and hormones to prevent diseases and grow disproportionally big. The end is always the same—death followed by the flesh being cut up, packaged, and shipped to supermarkets. One of the reasons for this is that to allow animals to live adequate lives, the cost of meat would be very high and the supply very low. Thus, it seems enticing the possibility of replacing factory-farmed animals with meat that cannot suffer.

But if the concern is the "plight of animals," it seems that what is in play here is a virtuous attitude that is not yet brought to fruition. If the very project of cultured meat is justified in terms of reduction of suffering, it is clear that the idea behind is that suffering is not okay, that is, that there is something inherently callous about allowing animals to suffer. Consequently, it would seem that the best solution is to be consistent in the application of compassion and make a serious effort to do away with factory farming altogether and educate the public, and ween them, out of animal products.

If the underlying concern is animal suffering, it just seems callous, greedy, and self-indulgent to eat meat in the first place. Considering that meat is not a requirement for good health, in fact, quite to the contrary, science shows that animal products can be harmful to human health;[24] and considering that taste can easily be adjusted to plant food, rather than perpetrating the idea of meat, our efforts as a civilization should be pursuing ways to move toward a plant-based diet. Is it possible that we have made such a mess of things in the world by bringing into existence millions of animals for food resulting in the irreversible degradation of the environment and of our health that we now are contemplating eating lab-made food? It seems to me that cloning meat is just another step toward alienation from nature. As Z. Bhat, S. Kumar and H. Fayaz point out,

> Another problem with the in vitro meat production system is that it may alienate us from nature and animals and can be a step in our retreat from nature to live in cities. Cultured meat fits in with an increasing dependence on technology, and the worry is that this comes with an ever greater estrangement from nature. In the absence of livestock based farming, fewer areas of land will be affected by human activities which is good for nature but it may at the same time alienate us from nature.[25]

I think that the enthusiasm about lab-grown meat is mostly due to self-indulgence (as well as short-sightedness). We should ask ourselves whether the mere taste of a food is so important that we are willing to produce it in a laboratory. What I am referring to here is something along the lines of temperance, or lack thereof. Human beings are animals endowed with reason. As animals, they are naturally subject to appetites for food, drink, sex, and more. They are sensitive to the pleasures that the satisfaction of such appetites can bring. Since our animality is not the distinguishing aspect of our humanity, physical pleasures should not be of major importance to us. However, humans are susceptible to these pleasures because our animality is part of their essence. In other words, insofar as we are part animals and part rational beings, we have to deal with all kinds of physical pleasures in a way that harmonize with reason.

Temperate people relate properly to their animality, and give the proper worth to animalistic pleasures. Insensible people and self-indulgent or intemperate, in their respective way, misjudge the importance of certain pleasures and misjudge themselves. Intemperate people place too much importance to the pleasures of food and drink. We eat and drink, primarily, because we require nourishment. Thanks to the ease of modern civilization, people who live in affluent societies, in my view, have lost sight of this and have placed too much importance to food. Food nowadays, for those of us who live comfortably, is more than fuel for the body.

Food is necessary and pleasurable. But humans can (and indeed do) have the wrong desire for it; generally, in affluent societies, this is manifested by the excess of food that people eat. For example, according to the Office of Disease Prevention and Health Promotion's dietary guidelines 2015–2020, "The typical eating patterns currently consumed by many in the United States do not align with the Dietary Guidelines."[26] To say that eating patterns in the US "do not align with the dietary guidelines" is a very mild way to put it when we consider that, "About three-fourths of the population" consume a low amount of fruit and vegetables. Also, "More than half of the population is meeting or exceeding total grain and total protein foods recommendations, [and] are not meeting the recommendations for the subgroups within each of these food groups." In particular, "most Americans exceed the recommendations for added sugars, and saturated fats."[27] And saturated fats come mainly from animal sources, including meat and dairy products. Furthermore,

> the eating patterns of many are too high in calories.... The high percentage of the population that is overweight or obese suggests that many in the United States overconsume calories ... more than two-thirds of all adults and nearly one-third of all children and youth in the United States are either overweight or obese.[28]

These facts in my view clearly show that something about our relation with food has gone completely wrong. These facts are not surprising considering that the idea of food, for many reasons and by many factors, has been distorted.

It seems clear that it is the lack of temperance that makes humans take pleasure in the wrong food and in the wrong way. Furthermore, self-indulgence leads to pain, more than it is required, when certain foods are missed. The self-indulgent values food too highly, choosing it at the cost of health. Thus, in relation to the bodily pleasure of food, one can be either self-indulgent, weak-willed, self-controlled, temperate, or insensible. The temperate person will choose what is pleasant and conducive to health, which is in its turn conducive to flourishing. Consequently, since strong desires for food can easily lead us astray in terms of health, the temperate person desires simple food and in moderation. When we survey the health sciences, it is clear that the only foods that can lower and prevent many health problems are fruits and vegetables.[29] In fact, as far as I have researched, I have never seen any study, or heard any medical professional, warn us of eating fruit and vegetables with caution.

Lab-grown human meat

Synthesizing meat opens the door to the variety of meat that can be produced. It seems plausible that if laboratories crack the code and succeed in creating perfect replicas of meats, the next step would very likely be replicating the flesh of endangered species, wild animals, and humans. Cannibalism is not a desirable practice in modern society and not only because the very idea of it is repulsive but because it can cause a disease known as Kuru.[30] But what if human flesh can be replicated without the risk of any disease? Though human flesh is unlikely to become a popular dish, given the curiosity of human beings, there is still the prospect of cloning it for human consumption. This may sound like a slippery slope objection to lab-grown meat, but I don't think that it is in this case. I do not intend to pursue this as an argument against lab-grown meat, however, but rather use the discussion to illustrate the kind of irrational path to which cloning meat leads.

Owen Schaefer and Julian Savulescu point out that,

> The most obvious reaction to this possibility of human [in vitro meat] is to ban it. Just as, for instance, cloning is banned in the 13 US states and the European Union for moral reasons, we could put in place strict restrictions on the synthesis of human flesh for the purpose of consumption. Given common revulsion at the prospect of cannibalism, this reaction is indeed rather likely. However, it is too quick—we should ask first, what is so wrong with cannibalism of artificially created human cells and tissue that it must be banned?[31]

Here they point out that despite our gut feeling sense that cannibalism is wrong, in the end there is no good argument against eating human flesh as it merely amounts to feeling of disgust. In fact, cannibalism is morally objectionable because it (typically but not always) involves killing a person, and the desecration of a corpse. But if human flesh is cloned, then there is no killing or desecration involved. To produce in vitro human flesh for human consumption it would be required to harvest cells from people who are willing to donate their cells. In fact, this process may even become lucrative for many people who might be paid for their cells. At that point then, what would be wrong with eating human flesh?

Could it be disrespect toward humanity? Since there are no human beings required in the production of a hypothetical in vitro human flesh, no disrespect could be done. Schaefer and Savulescu thus conclude that if we are worried about in vitro meat because of cannibalism, we should not worry at all because such meat will be free of cruelty and disrespect. Therefore, the objections that are typically raised against lab-grown meat rely on violation of respect and disgust, but they are not strong enough to reject the project of cloning meat for human consumption. Fewer animals being slaughtered, less animal suffering, less pollution, among many other factors, in their view, are powerful enough arguments showing that we should support research into cultured meat.

Many other opinions about cloned meat have a similar attitude to that of Schaefer and Savulescu. Namely, as I already pointed out at the outset, it seems that there is a prevailing view about the moral viability of making lab-grown meat that hinges on broadly consequentialist and deontic principles. Virtue ethics is not necessarily against the best consequences or the notion of rights. However, those should not be the only aspects that matter. As we have seen in the discussion of Hursthouse about abortion, sometimes in the name of the best consequences or in the name of our rights, we might act in ways that are callous, self-indulgent, selfish, and so on. Thus, it would certainly be an admirable thing to reduce suffering and care for the environment. But is cloning meat the right way to accomplish those things? I would like to suggest what a virtue-oriented approach can add to the discussion, and in so doing I will address cannibalism in particular, though the larger scope is to address in vitro meat. What I would like to suggest is something along the lines of what Leon Kass refers to as "the wisdom of repugnance," which is the same concept that Midgley and others calls the "yuck factor." This is the notion that a strong, negative reaction of disgust to a practice is in fact good enough evidence that such a practice is not morally sound or that there is something intrinsically wrong with it. Granted, this sounds just like an appeal to emotion fallacy. The appeal to emotion fallacy, it has to be remembered, is an informal fallacy, which means that it is possible for the conclusion of an argument based on such premises to be true. Martha Nussbaum points out that the "yuck factor" or disgust has been used in many arguments throughout history as a justification for evil practices and institutions, such as slavery, torture, antisemitism, gender and sexual discrimination, and so on.[32] But it seems to me, and many others, that just because feeling of disgust may lead to the wrong conclusion, it does not follow that this feeling should be discounted forthright. Our feeling of revulsion may not be in itself an argument against a practice, but it certainly signals that something requires our attention because it might be morally wrong. Surely, we can in many cases supply reason to this feeling and construct an argument. But even in the case that a fully articulated argument is not forthcoming, I don't think that in certain cases one is not entitled to reject a practice, like in this case in vitro meat, on the basis of disgust. In fact, Leon Kass seems to think so, as he argues the following about the feeling of revulsion,

> Revulsion is not an argument; and some of yesterday's repugnances are today calmly accepted—though, one must add, not always for the better. In crucial cases, however, repugnance is the emotional expression of deep wisdom, beyond reason's power fully to articulate it. Can anyone really give an argument fully adequate to the horror which is father-daughter incest (even with consent), or having sex with animals, or mutilating a corpse, or eating human flesh, or even just (just!) raping or murdering another human being? Would anybody's failure to give full rational justification for his or her revulsion at these practices make that revulsion ethically suspect? Not at all. On the contrary, we are suspicious of those who think that they can rationalize away our horror, say, by trying to explain the enormity of incest with arguments only about the genetic risks of inbreeding.[33]

The natural feeling of repugnance at cannibalism, and in general at cultured meat, belongs in this category. We are repelled by the prospect of cannibalism and cloned meat because we feel directly that such a practice violates our moral virtues. What kind of person am I to support artificial meat when I thrive by eating plants and fruit? Is the taste of meat so important that we are willing to allow technology to take over our lives? These are some of the questions that are part of that feeling of disgust. Repugnance, thus, is a natural reaction against the excesses of human willfulness to distance itself from nature. In this case, I believe, the repugnance expressed at the prospect of producing meat artificially is justified. In the case of sexism, racism, slavery, and more, repugnance stems from contempt, anger, and self-delusion. Thus, nurturing and following that feeling is wrong. But in the case of cultured meat, repugnance is the cry out of our human nature that is being overtaken and changed by technology, the blind hunger for innovation, profit, and self-indulgence.

As a concluding remark, I would like to point out that my discussion about the morality of producing and eating lab-grown meat is supposed to illustrate what a virtue-oriented approach can add to the discussion. Moreover, it is a view that the ethical vegan might take with respect to the morality of in vitro meat. Ethical veganism, as I understand it, is the rejection of meat as food, whether it is from an animal or a lab. What I argue is that ethical veganism should be based on virtue rather than deontic or consequentialist principles.[34] Ethical veganism should be the embodiment of virtue and thus should reject the notion of using animals for our taste and pleasure because doing so evinces a lack of temperance, compassion, fairness, and magnanimity. Consequently, an ethical vegan should not support the production of any kind of meat. However, not all vegans think this way. For example, Ingrid Newkirk, founder and president of People for the Ethical Treatment of Animals (PETA), offered $1 million to successful production of lab-grown meat.[35] As an ethical vegan, this seems to me a peculiar form of veganism. In my view, the reason for this schism among vegans is due to the fact that the question of the morality of lab-grown meat has been framed typically in terms of potential future gains, which seems to me to be an approach of consequentialist nature; or another typical approach is of deontic nature. These approaches seek to rationalize the question of lab-grown meat, and certainly in their way they seem to achieve the goal of demonstrating that cloning meat for human consumption is morally viable and makes a lot of sense to support the project. After all, isn't less pollution and less animal suffering what we all want?

Yes, but as I hope to have shown, while those are important factors, they are not the only factors to be considered. Focusing only on those factors may lead to a tunnel vision understanding of the issue. The contribution of a virtue-oriented approach is to show that, for example, the way we are going about reducing suffering and environmental degradation seems to completely disregard the importance of having an admirable character. In a recent article, Jean-François Hocquette aptly concludes,

> the global scientific community including the proponents of artificial meat themselves recognize the hurdles to overcome so that artificial meat can progress to the industrial stage (new formulation of culture media, development of giant incubators, safety assessment for human consumption, etc.).[36]

He also notes that there are other alternatives to cultured meat that "faster to develop in the short term and more effective in responding to today's issues (in particular it is the case of the reduction of waste) compared to artificial meat which still needs a great deal of research."[37] A viable solution is, of course, that of plant-based meat substitutes, though in my view this would reinforce the idea that humans need meat or, if no meat is around, anything that looks and smells as close as possible like meat. It would be more sustainable in the long run if we take steps toward abandoning what I regard as a primitive idea of animals as human food and we embrace the human diet.

Notes

1 Many terms are used to denote synthesized meat: lab-grown, test-tube meat, clean meat, synthetic meat, in vitro meat, and more.
2 Gretchen Vogel, "Organs made to order." *Smithsonian Magazine*, 2010. www.smithsonianmag.com/science-nature/organs-made-to-order-863675/
3 Johns Hopkins Bloomberg School of Public Health, "Health & environmental implications of U.S. meat consumption & production." www.jhsph.edu/resea

rch/centers-and-institutes/johns-hopkins-center-for-a-livable-future/projects/meatless_monday/resources/meat_consumption.html
4 "The Einstein God," *Reddit*. www.reddit.com/user/the_einsteinian_god
5 Ibid.
6 Arielle Duhaime-Ross, "Test-tube burger: Lab-cultured meat passes taste test (sort of)." www.scientificamerican.com/article/test-tube-burger-lab-culture/ Also "World's first lab-grown burger is eaten in London." www.bbc.com/news/science-environment-23576143
7 "Livestock and Climate Change." Worldwatch Institute, www.worldwatch.org/node/6294
8 The following is a list of sources: Mario Herrero et al., "Biomass use, production, feed efficiencies, and greenhouse gas emissions from global livestock systems." *Proceedings of the National Academy of Sciences*, 110(52), 2013: 20888–20893. doi:10.1073/pnas.1308149110 ; Richard Oppenlander, "Freshwater abuse and loss: Where is it all going?" 2013. www.forksoverknives.com/freshwater-abuse-and-loss-where-is-it-all-going/#gs.xz97at; Mesfin M. Mekonnen and Arjen Y. Hoekstra, "A Global Assessment of the Water Footprint of Farm Animal Products." *Ecosystems*, 15(3), 2012: 401–415. doi:10.1007/s10021-011-9517-8; P. W. Gerbens-Leenes, M. M. Mekonnen, and A. Y. Hoekstra, "The water footprint of poultry, pork and beef: A comparative study in different countries and production systems," *Water Resources and Industry, Water Footprint Assessment (WFA) for Better Water Governance and Sustainable Development*, 1–2, 2013: 25–36. doi:10.1016/j.wri.2013.03.001 ; Pete Smith, Mercedes Bustamante et al., "Agriculture, Forestry and Other Land Use (AFOLU)." www.ipcc.ch/site/assets/uploads/2018/02/ipcc_wg3_ar5_chapter11.pdf ; Philip Thornton, Mario Herrero, and Polly Ericksen, "Livestock and Climate Change." *International Livestock Research Institute*, 2011. https://cgspace.cgiar.org/bitstream/handle/10568/10601/IssueBrief3.pdf; Richard Oppenlander, *Food Choice and Sustainability: Why Buying Local, Eating Less Meat, and Taking Baby Steps Won't Work* Minneapolis, MN: Langdon Street Press, 2013; Sérgio Margulis, "Causes of deforestation of the Brazilian Amazon." World Bank Working Paper, no. 22. Washington, DC: World Bank, 2004;
Louisiana Universities Marine Consortium. "What causes ocean 'dead zones'?" *Scientific American*, 2016; Environmental Protection Agency. "What's the problem? Animal Waste Region 9 US EPA." https://www3.epa.gov/region9/water/; Henning Steinfeld et al., "Livestock's Long Shadow." World Wildlife Fund. www.europarl.europa.eu/climatechange/doc/FAO%20report%20executive%20summary.pdf; Center for Biological Diversity. "How *Eating Meat Hurts Wildlife and the Planet.*" Take Extinction off Your Plate (n.d.).
9 Hanna L. Tuomisto, and M. Joost Teixeira de Mattos, "Environmental impacts of cultured meat production," *Environmental Science & Technology* 45(14) (July 15, 2011): 6117–23, doi:10.1021/es200130u.
10 Henceforth, I will use the terms lab-grown meat or cultured meat or other such equivalent terms interchangeably.
11 Alok Jha, "Synthetic meat: How the world's costliest burger made it on to the plate," *The Guardian*, August 5, 2013; Nick Collins, "Test tube hamburgers to be served this year." *The Telegraph*, February 19, 2012; Carolyn S. Mattick et al., "Anticipatory life cycle analysis of in vitro biomass cultivation for cultured meat production in the United States." *Environmental Science & Technology*, 49(9), 2015: 11941–11949; Carolyn Mattick et al., "The problem with making meat in a factory," *Slate*, 2015. https://slate.com/technology/2015/09/in-vitro-meat- probably-wont-save-the-planet-yet.html
12 Ibid.
13 Collins, 2012.

14 "Why cultured meat." http://whyculturedmeat.org/essays/animal-rights/is-it-animal-rights/; Notaro, K. "The crusade for a cultured alternative to animal meat: An interview with Nicholas Genovese, PhD PETA." 2011. https://ieet.org/index.php/IEET2/more/notaro20111005; "New Harvest – FAQ." New Harvest. http://whyculturedmeat.org/faq/
15 Carlo E. A. Jochems et al., "The use of fetal bovine serum: Ethical or scientific problem?" *Atla-Nottingham*, 30(2), 2002: 219–228.
16 Ibid.
17 Betti M. I. Datar, "Possibilities for an in Vitro Meat Production System." *Innovative Food Science & Emerging Technologies*, 1, 2010: 13–22, doi:10.1016/j.ifset.2009.10.007
18 Charlotte Hawks, "How close are we to a hamburger grown in a lab?" *CNN*. www.cnn.com/2018/03/01/health/clean-in-vitro-meat-food/index.html
19 M. Specter, "Test-tube burgers." *The New Yorker*. www.newyorker.com/magazine/2011/05/23/test-tube-burgers
20 D. M. Fessler et al., "Disgust sensitivity and meat consumption: A test of an emotivist account of moral vegetarianism." *Appetite*, 41, 2003: 31–41 ; Diana Fleischman, "Lab Meat: Survey Results." *The Vegan Option Radio Show and Blog*, May 16, 2012. https://theveganoption.org/2012/05/16/lab-meat-survey-results/; Wim Verbeke et al., "Would You Eat Cultured Meat? Consumers Reactions and Attitude Formation in Belgium, Portugal and the United Kingdom." *Meat Sci.*, 102, 2015: 49–58. doi:10.1016/j.meatsci.2014.11.013.
21 Carolyn Mattick, and Brad Allenby, "The future of meat: Issues in science and technology." *Issues in Science and Technology*, 30(1), 2013.
22 Jasmijn de Boo, "The future of food, why lab grown meat is not the solution." *Huffpost*, September 10, 2013. www.huffingtonpost.co.uk/jasmijn-de-boo/lab-grown-meat_b_3730367.html
23 See many arguments offered in favor of cultured meat by Why Cultured Meat, http://whyculturedmeat.org/essays/animal-rights/is-it-animal-rights/
24 University of Southern California. "Meat and cheese may be as bad for you as smoking." *ScienceDaily*. www.sciencedaily.com/releases/2014/03/140304125639.htm
25 Z. Bhat, S. Kumar and H. Fayaz, "In vitro meat production: Challenges and benefits over conventional meat production." *Journal of Integrative Agriculture*, 14, 2014: 241. doi:10.1016/s2095-3119(14)60887-x
26 https://health.gov/dietaryguidelines/2015/guidelines/chapter-2/current-eating-patterns-in-the-united-states/
27 Ibid.
28 Ibid.
29 www.hsph.harvard.edu/nutritionsource/what-should-you-eat/vegetables-and-fruits/
30 D. C. Gajdusek and V. Zigas, "Degenerative disease of the central nervous system in New Guinea. The endemic occurrence of "kuru" in the native population." *New England Journal of Medicine*, 257, 1957: 974–978.
31 G. O. Schaefer and J. Savulescu, (2014). "The ethics of producing in vitro meat." *Journal of Applied Philosophy*, 31(2), 2014: 188–202. doi:10.1111/japp.12056
32 Martha Nussbaum, "Danger to human dignity: The revival of disgust and shame in the law." *The Chronicle of Higher Education*, 50(B6), 2004.
33 Kass, Leo. "The wisdom of repugnance." *The New Republic*, 216(22), 1997, p. 20.
34 Carlo Alvaro, "Ethical veganism, virtue, and greatness of the soul." *J. Agric. Environ. Ethics*, 30, 2017: 765. doi:10.1007/s10806-017-9698-z; Carlo Alvaro, "Veganism as a Virtue: How compassion and fairness show us what is virtuous about veganism." *Future of Food: Journal of Food, Agriculture and Society*, 5(2), 2017.
35 https://abcnews.go.com/Technology/story?id=4704447&page=1.
36 J. F. Hocquette, "Is in vitro meat the solution for the future?" *Meat Science*, 120, 2016: 8.
37 Ibid.: 9.

4 Ethical veganism
What it is, what it is not, and what it should be

Veganism today has become an ambiguous term. When I decided to become a vegan the first time back in the 1980s, the term vegetarian was used to denote a person who eats only vegetables; that is, a vegan. Vegetarianism today means something different. It is used to describe a person who does not eat any kind of meat, but consumes dairy products and other animal byproducts. The reasons for choosing vegetarianism vary. There are vegetarians who believe that eating meat is morally wrong but it is not wrong, for example, to use leather and eat cheese. Many vegetarians, however, are not interested in animal ethics; they do not eat meat just because they don't like the taste of it. Still others do not have any moral position. And many others are dietarily flexible, which means that they eat anything depending on the occasion or other factors (or anything that moves, some might say).

What is veganism?

Veganism, as I said, is an ambiguous term, and a ubiquitous term, too. Like anything else capable of generating profit, veganism has been distorted and twisted into many different forms because the label "veganism" is fashionable and very lucrative these days. In fact, veganism, which started as a revolutionary philosophy, is no longer revolutionary. It has been swallowed by corporations and has become one more entry in a long list of fad diets. As a recent article points out, veganism seems to be more helpful for companies to increase their profits than it is to the environment.[1] Granted, there still are vegans who believe in veganism as an important moral philosophy. But what exactly is veganism? One very broad way to put veganism is the Vegan Society's definition:

> Veganism is a way of living which seeks to exclude, as far as is possible and practicable, all forms of exploitation of, and cruelty to, animals for food, clothing or any other purpose.[2]

This definition seems clear enough, though it is incomplete. Why incomplete? For one thing, such a definition does not state why one seeks to exclude, as far as is possible and practicable, animal exploitation and consumption of animal food. It says, "seeks to exclude." Does this mean that one must seek to exclude or ought to seek to exclude or ... In other words, such a statement seems to be intended as a moral statement; however, it does not specify on which moral ground one seeks to exclude animal exploitation and consumption of animal food. Second, the statement is incomplete in that it does not specify criteria to determine how much is "as far as is possible." Perhaps what the definition means is that there are circumstances where avoiding animal products is impossible. For example, the glue used in Band-Aids and the lubricant used in razor heads are animal-derived. Many household products contain animal by-products or are tested on animals. Animal fat is used in the production of lotions, glue, paint, motor oil, and more. Vanilla, raspberry and strawberry ice cream flavors are enhanced by castoreum, the secretion from the castor sacs of the beaver! Some brands of chewing gum contain lanolin, an oily, sweaty secretion found on the outside of sheep's wool. Lanolin is also used in skin lotions and laundry softeners.

Candies are colored using the secretions of various insect or crushed bugs. There are animal parts and animal by-products everywhere—in soap, detergent, cars, candies, animals, animals, and more animals. In other words, there are animal by-products in many ordinary items that we use daily. Worse, in many cases there is no way to know whether a product is truly vegan. For example, this very computer that I am using to type these words and the glue they used to glue my shoes might contain animal by-products. So what is veganism? Veganism is a social and political philosophy that shuns the practice of using animals for food and other purposes; in other words, veganism rejects animal exploitation.

Plant-based diets

There is another equally ubiquitous term: "plant-based." Some people use this term to mean that they are vegans. Then why don't they just use the term vegan? There are many psychological as well as practical reasons for this. A psychological reason is that, since many consider veganism as an extreme diet, the term "plant-based" conveys a more flexible message about diet than veganism, giving the feeling that one is not required to follow a strict regimen or discipline. Another reason is that some people are put off by ethics. Veganism is a specific ethical position with respect to animals, while "plant-based" refers not to morality but to diet. Another reason is that over the years the term vegan has acquired a less-than-positive connotation. For example, veganism is associated with PETA, which is a controversial organization. I say controversial because of their somewhat strange views on animals. For example, they declared that what human beings do to animals is comparable to the Holocaust. I happen to agree with that statement, although I would not necessarily make such a statement or suggest anyone making that statement. Also they said that there would be nothing wrong with consuming eggs from companion chickens: "We would not oppose eating eggs from chickens treated as companions if the birds receive excellent care and are not purchased from hatcheries."[3] In other words they say that if one takes care of chickens, and treats them like family members, it would be okay to use their eggs as food. Now this statement I don't agree with. If PETA is a vegan organization, why are they even telling people that it is okay, in certain circumstances, to eat eggs? Then, would it be right to eat meat as long as an animal is treated like a family member? Obviously, PETA may retort that no one eats a family member. But what about eggs? Why is okay to eat the eggs of a family member? Why is it okay to eat something that comes out of the rear end of an animal, family member, human, or non-human? Furthermore, PETA's president, Ingrid Newkirk, is on the record opposing companion animals and supports the euthanizing and extermination of all pit bulls. Thus, many people feel that the term "vegans" fits too tight and prefer the term "plant-based."

What veganism is not

Veganism is not specifically a diet (although it is possible to talk about vegan diets), but rather a moral lifestyle. Moreover, veganism is not a cure to diseases and it does not have weight loss properties per se. What I mean here is simply that excluding animal-based food from one's diet is a great idea from a medical and an ethical point of view. However, many people expect that going vegan will automatically cause them to lose weight, improve physical and mental performances, and more. This is a simple point, but a misunderstood point, as well. There are many vegan diets, some of which are not conducive to good health. As I argue in this book, a raw vegan diet of mostly raw fruit, tender leafy greens, and occasionally nuts and seeds, is the optimal human diet. As cooked food is incorporated in the human diet, a diet becomes less and less conducive to good health.

Veganism is the strict sense is an ethical position that opposes animal exploitation; this includes shunning all insect-based and animal-based food, such as meat, eggs, dairy products, honey, and more, and oppose the use of products derived from animals, such as fur, leather, body care products and cosmetics that contain animal by-products. Furthermore, in many cases vegans are opposed to circuses, and other forms of animal entertainment, and even owning companion animals. However, individuals who identify themselves as vegans are not like members of a religious denomination. That is to say, vegans approach veganism differently. Some vegans, for example, are not against owning companion animals, but some are. And with regard to food, while veganism is supposed to be a position that advocates the avoidance of animal food, many who identify as vegans do not mind eating meat in particular occasions, such as holidays, for example. Certainly, this behavior is not typical; some might even consider it as hypocritical. Furthermore, some vegans may eat mollusks or honey, and may have no compunction about wearing leather shoes or clothes made of wool or silk. Thus, it is important to take these facts into consideration when we read studies that purport to show the benefits or drawbacks of certain diets.

Considering the foregoing, it is important to understand that veganism is not one single, defined diet, and thus it is not necessarily a healthful diet. For example, fried potato chips, dairy-free pastries, plant-based cheeses, beer, cigarettes, heroin, and mock meats, and more, are vegan. Virtually all experts in nutrition deem such examples of processed foods, and drugs, unhealthful. Consequently, research does not show that all vegan diets can benefit our health. Rather, a vegan diet is beneficial when it lacks the presence of animal products as well as the lack of highly processed and refined food; in fact, as I argue the only healthful form of veganism is the human diet, which is a diet of raw and fresh leafy greens and fruit, with the occasional nut and seed. As I will discuss later, there are different levels of healthful nutrition, that is, there is optimal nutrition and non-optimal. Many non-optimal diets are still superior to eating animal foods. But others are perhaps inferior.

In the past, a vegan would consume vegetables, grains, legumes, nuts, seeds, and perhaps on occasion soya byproducts. Nowadays there are a plethora of vegan alternatives—in short, for every conventional food one can think of, nowadays there exist a vegan version, from vegan cheese to all kinds of mock meats. On the one hand, these products spice up the life of the vegan offering variety; on the other, most are unhealthful. Thus, a vegan is not synonymous with optimal health. Considering that foods labeled as vegan can be high in fat, salt, refined sugar, and often low in key nutrients, it is not surprising that poor choice can lead vegans to have health problems. What emerges from scientific research, which also makes logical sense, is that vegan diets are conducive to good health when they exclude or considerably minimize processed foods.

Misconceptions about veganism

Also, there are many people who become vegans specifically for health reasons. While it is true that veganism can be conducive to good health, it is not necessarily true. In fact, as I stated above, veganism is not a diet, but an ethical lifestyle. Being a vegan does not tell much about one's diet. For example, Oreos, French fries, coffee, alcohol, and other kinds of junk food, are not animal-based products. So, one who consumes Oreos, French fries, coffee, and alcohol daily is by definition a vegan. However, if one's diet consisted only in Oreos, French fries, coffee, and alcohol, one would very soon become very sick. Many people do not understand this point and so they have many misconceptions about veganism. Of course, being a vegan like my example above is dangerous. But veganism itself can be very beneficial.

I want to talk about some of the common misconceptions that people have about veganism. When I talk to people, especially to meat eaters, they tell me that eating meat is essential for good health and that veganism is unhealthful. I must say that the scientific evidence shows the opposite: eating animal products causes many health problems. The World Health Organization, for example, has shown that eating certain kinds of meat, such as red meat and processed meat, i.e., ham, sausages, pastrami, is likely to cause of cancer.[4] About the now oft-cited WHO study, it has to be clarified that the study does not say that meat equals cancer. The study does say that processed meats, such as hot dogs, bacon, and cold cuts, are carcinogenic. Regarding red meat, they state that it is likely carcinogenic. The study also found no particular connection between other types of meat and cancer. However, it has to be borne in mind that the WHO as well as most other health organizations warn the public to consume meat and animal products in moderation and fruits and vegetables in abundance. My point is that disputes aside, animal products are not the type of foods that are recommended in abundance, and this is a very telling story. The most sensible explanation is that animal-based food is not very healthful. Also, according to a scientific study, consuming cow's milk can cause diabetes.[5]

Here are some of the critical nutrients that meat eaters typically worry about.

Proteins

Whether it is in meat or vegetables, proteins are the same. There are no special proteins contained only in meat. Especially meat eaters (those who are ignorant about nutrition) like to point out, "If you do not eat meat, where do you get your proteins?" Obviously, those who ask this question ignore the fact that proteins are present in all living organisms. So, you can get your proteins by eating a raw diet of watermelon, lettuce, dates, and many other fruits, greens, nuts, and seeds. The point is that proteins are complex structures composed of amino acids.[6] There are in total 20 different amino acids. The human body synthesizes 11 of them, while the other nine must be obtained through nutrition.[7] These are known as 'essential amino acids'.[8] Not surprisingly, since human beings are frugivores, these nine essential amino acids are found in leafy greens and fruits in optimal quantities.[9] Most importantly, it is not necessary to consume a single food that contains a variety of amino acids. As long as one consumes a variety of fruits and leafy greens it is quite simple to obtain all required amino acids.[10] Thus the question of where vegans get their proteins is not an issue at all. If one eats plenty of fruits and vegetables, he or she will have an optimal amount of proteins. In fact, in a recent scientific article, the American Dietetic Association states, "appropriately planned vegetarian diets, including total vegetarian or vegan diets, are healthful, nutritionally adequate, and may provide health benefits in the prevention and treatment of certain diseases." Moreover, such diets are beneficial to all individuals, from infants to athletes.[11]

Vitamin B12

Another point of contention is vitamin B12. I discuss this topic in Chapter 7. However, I shall state a few points here. First of all, animal products do not have magical powers, so to say, of containing B12. Animals must acquire the vitamin just like humans do. If the only way to obtain B12 were through the consumption of animal products, then, herbivore animals, not eating animal products, would be vitamin B12 deficient. Second, people who consume animal-based diets can experience B12 deficiency due to poor absorption. For example, deficiency in iodine may cause B12 deficiency; and one could eat as much animal-based food as he cares to and still experience low B12. Furthermore, there are other factors that cause low B12, such as alcohol, tobacco, drugs, antibiotics, and many other (typically cooked) foods and products that destroy the bacteria responsible for the production of B12.

So, from which source do animals obtain B12? The short answer is that it is produced by bacteria and synthesized by bacteria in the mouth and the gut flora in humans and animals.[12] In order for bacteria to produce B12, the soil needs to contain enough cobalt. However, due to declining soil quality as a result of environmental degradation, nowadays the soil is deficient in cobalt; consequently, acquisition of B12 is a problem for both animals and humans. Early humans would obtain plenty of B12 from drinking uncontaminated water, fruit, and leaves. As I explain in Chapter 7, due to the declining soil quality, farm animals are given B12 supplementation. In fact, more than 90% of all B12 supplements manufactured are given to farmed animals. Thus, people who consume animal-based diets obtain B12 from the supplements fed to the animals. Raw vegans can easily cut out the middleman and, if necessary, can obtain B12 by taking supplements to meet this need.

Omega-3 fatty acids

Omega-3 fatty acids are very important to us because they help the heart beat at a steady rate, they lower blood pressure, and other important functions. Typically, we expect to find these omega fatty acids in seafood. So, the question is, since vegans do not eat seafood 'From where do they get omega-3?" This, again, is not a real issue; Vegetables and fruit contain omega-3s. Fish are not some sort of magical beings that ooze out omega fatty acids. They cannot make their own, but obtain them from algae. Thus, eating fish is not the only food that provides with omega-3 fatty acids. The fact is that fruit and vegetables contain optimal amounts of omega-3s. If this were not the case, prior to cooking and eating fish, our ancestors would have died. An interesting article by Harvard Medical School has the following title: "Omega-3 fatty acids: Does your diet deliver? Most Americans don't get the recommended amount of these potentially heart-protecting fats."[13] It is interesting because it is based on scientific research on "most Americans" and not most vegans! Since it is claimed that animal products abound with omega-3s (as well as other nutrients) one would not expect to find "most Americans" to be omega-3 deficient. The explanation is very simple: eating tons of seafood and meat and other animal products with the hope to get the recommended amounts of omega-3 is hopeless. Also, we have to keep in mind that the scientists do not exactly know the amounts of nutrients that one should get. In most cases the right amount is always lower than the recommended.

The problem, I argue, is cooked food. Eating food that has been either cooked or processed and made less bio-available is the problem. Although there aren't many, there are individuals with many decades of living on a raw vegan diet. If omega-3 were a problem for these individuals, they would be dead or very ill and by now the whole world would have known about them. All fresh and unprocessed foods contain essential fatty acids. More than 50% of the fat in greens like romaine lettuce and spinach is from omega-3s. Ten cups of these greens a day will supply an optimal amount of fatty acids. Furthermore, berries, guava, cherries, cantaloupe, honeydew, kiwi, papaya, mangoes, grapes, lemon, and more are great sources of omega-3s. For example, one medium cantaloupe provides a third of the recommended daily intake for omega-3 fats. Add that to other fruits and greens and it the omega-3s and 6s will abound. Repetition is necessary, thus I will make this point several times in this book: the human diet is a complete and optimal diet for our species. This consists in a large amount of fruit and tender leafy greens, which are rich in omega-3s, and the occasional bunch of nuts and seeds, which contain omega-6s.

Vitamin D

Vitamin D is very important because it promotes the absorption of calcium and bone growth. Accordingly, vitamin D "is a fat-soluble vitamin that is naturally present in very few foods… it is also produced when ultraviolet rays from sunlight strike the skin and trigger vitamin D synthesis."[14] While it is true that dairy products contain vitamin D, humans do not need to consume dairies. As I explain in Chapter 7, there is no research that indicated that vitamin D deficiency is endemic among vegans or raw vegans.

The public has many misconceptions about veganism. But understanding easy concepts about the science of nutrition and the human species can correct these misconceptions. Veganism can be very beneficial to our health; however, we should not assume that not eating meat or animal products is automatically beneficial to our health.

What it should be

Veganism should be an expression of virtue. In previous chapters I focused on the virtues of nonviolence and aesthetic. Veganism should be understood as the embodiment of the virtue of moderation. To put it in argument form,

1 Animal-based diets are not essential for good health, they are the leading cause of environmental degradation, produce un-aesthetic values and violence (as discussed in Chapter 2), and can cause a number of chronic diseases.
2 It follows, then, that consuming animal products is based solely on unregulated gustatory values or practicality.

3 Moderation as a virtue of character is conducive to a good moral life. For example, one who is typically moderate acquires an internal harmony of peace within himself for not wanting too much or too little.
4 Moderation in one's diet gives one an internal harmony; moreover, it is a rational regulation of one's appetites, which is conducive to good health.
5 Thus, a moderate person avoids consuming animal products altogether because there is no moderate way to do it in the light of the considerations made in premise 1).
6 Therefore, it must be concluded that a moderate person embraces a vegan diet.

So what should veganism be, again? If one has to take one thing from this chapter it would have to be this: Veganism nowadays appears to be more concerned about veganism than about animals. Veganism is foodism, a lifestyle that promotes preparation and enjoyment of food. "What is wrong with that?" one may ask. In general, there is nothing wrong with enjoying what you eat. But I have two points: One is that humanity is drifting farther and farther away from its natural ways. Human beings are too dependent on technology. As we have seen in the previous chapter, we are so dependent on technology that instead of returning to nature, we are trying to produce lab-grown animal products. Most capitalistic societies have become a caricature of humanity where people consume big and eat more food that they need. I don't believe I need to show any scientific research to that effect. Just look around and see the staggering number of overweight people in affluent societies. Second, veganism should not be focused on what to eat next and how to make plant-derived proteins look like or taste like some animal products. Veganism should be an expression of virtue, a return to simplicity, and a way back to our nature—a way to listen to our body and realize that our food is a raw diet of fruit and tender leafy greens. A raw vegan diet of mostly fruit, some tender leafy greens, and the occasional nuts and seeds, is what veganism should look like. This is, what I will discuss in the next chapter, the diet that is best and optimal for humans.

Notes

1 Marryn Somerset Webb, "The veganism boom does more for food company profits than the planet." *Financial Times*, February 22, 2019. www.ft.com/content/79c2aa 16-35f1-11e9-bb0c-42459962a812
2 The Vegan Society. "Definition of veganism." www.vegansociety.com/go-vegan/ definition-veganism.
3 PETA. "Is it OK to eat eggs from chickens I've raised in my backyard?" www.peta. org/about-peta/faq/is-it-ok-to-eat-eggs-from-chickens-ive-raised-in-my-backyard/
4 V. Bouvard, D. Loomis, K. Z. Guyton et al., "Carcinogenicity of consumption of red and processed meat." *The Lancet Oncology*, 16, 2015: 1599. doi:10.1016/s1470-2045(15)00444–00441
5 T. Saukkonen, S. Virtanen, M. Karppinen et al., "Significance of cow's milk protein antibodies as risk factor for childhood IDDM: Interactions with dietary cow's milk

intake and HLA-DQB1 genotype." *Diabetologia*, 41, 1998: 72. doi:10.1007/s001250050869
6 "What are proteins and what do they do?" U.S. National Library of Medicine, 2019. https://ghr.nlm.nih.gov/primer/howgeneswork/protein
7 National Research Council (US). *Subcommittee on the Tenth Edition of the Recommended Dietary Allowances. Recommended Dietary Allowances*, 10th edn. Washington, DC: National Academies Press, 1989. www.ncbi.nlm.nih.gov/books/NBK234922/
8 "Amino acids." U.S. National Library of Medicine. MedlinePlus (n.d.). https://medlineplus.gov/ency/article/002222.htm
9 H. Ito, H. Ueno and H. Kikuzaki, "Free amino acid compositions for fruits." *J. Nutr. Diet. Pract.*, 1, 2017: 1–5.
10 R. H. Liu, "Health benefits of fruit and vegetables are from additive and synergistic combinations of phytochemicals." *Am J Clin Nutr*, 78, 2003: 517S-520S.
11 W. J. Craig, and A. R. Mangels, "Position of the American Dietetic Association: Vegetarian diets." *American Dietetic Association*, 109(7), 2009: 1266–1282.
12 M. J. Albert, V.I. Mathan and S. J. Baker, "Vitamin B12 synthesis by human small intestinal bacteria." *Nature* 283, 1980: 781–782.
13 "Omega-3 fatty acids: Does your diet deliver? Most Americans don't get the recommended amount of these potentially heart-protecting fats." Harvard Medical School, 2016. www.health.harvard.edu/heart-health/omega-3-fatty-acids-does-your-diet-deliver
14 "Vitamin D." National Institutes of Health, 2019. https://ods.od.nih.gov/factsheets/VitaminD-HealthProfessional/

5 Raw veganism
The human diet

Many environmental movements, such as the organic food movement, back to nature movements, nudism, sexual freedom, as well as raw foodism, have their roots in the Lebensreform (life reform) movement of the late nineteenth century.[1] One individual in particular, with the name of Maximilian Bircher-Benner (1867–1939), is credited to have started raw veganism. In the early twentieth century in America, cigarette smoking and alcohol were prevalent habits.[2] Bircher-Benner's idea was that since modern civilization has corrupted and alienated human beings from nature, living well requires a return to natural ways of living, a life devoid of stimulants, drugs, and processed food. In order to live well one must exercise, spend time outdoors in nature, and eat healthfully. Bircher-Benner thought that the ideal food for humans is plant food because plants absorb the sunrays, which are the primary source of vitality. Consequently, he adopted a vegan diet at first, and eventually came to the logical conclusion that raw foodism was the best possible way to eat.[3] After all, human beings are animals, and animals do not cook their food. An immediate objection here is that studies that have specifically argued for raw vegan diets are scarce. The mainstream view is that cooking in general, and the preparation of foods from animal origins in particular, have led to the expansion of the human brain, and the widely held view that dense foods and therefore also cooked foods may be required to maintain good health in light of our large brains.

Out of all the species of animals, humans are the only species that cooks their food—and the only ones who suffer from obesity, cardiovascular disease, atherosclerosis, diabetes, and more. Humans are primates. Biped beings were around over four million years ago, though large and more complex brains, language, and technological abilities developed much later, mainly during the past 100,000–200,000 years. Physical and genetic similarities indicate that *Homo sapiens* is closely related to the great apes. Humans and the great apes have a common ancestor that lived between eight and six million years ago. According to fossil records, humans first evolved in Africa between six and two million years ago. *Homo erectus* emerged about 1.8 million years ago, and *Homo sapiens* evolved 300,000 years ago in Africa.[4] Yet, the practice of cooking started 20,000 years ago.[5] Presumably, in that period humans prevalently consumed raw fruits and tender leafy greens. This is a very speculative area or research because it concerns a period of time so remote that it precludes researchers from clearly understanding what was in fact the case. However, using common sense, it is clear that prior to fire control and any sort of technology, humans ate fruits and tender leafy greens, and perhaps some insects.

Some suggest that cooking food contributed significantly to human's higher cognitive capacities. In an article in the *Proceedings of the National Academy of Sciences*, researchers Suzana Herculano-Houzel and Karina Fonseca-Azevedo argue that cooking food was the key factor to human increased encephalization:

> Absent the requirement to spend most available hours of the day feeding, the combination of newly freed time and a large number of brain neurons affordable on a cooked diet may thus have been a major positive driving force to the rapid increased in brain size in human evolution.[6]

Another key sources in the debate over the role of diet in human brain development is the expensive-tissue hypothesis proposed by Aiello and Wheeler. They explain that *Homo sapiens* developed a larger brain compared to other mammals has to be viewed as a concomitant event to the reduction in size of the gastro intestinal tract. According to the authors, this change was favored by "the incorporation of increasingly greater amounts of animal products into the diet [of *Homo sapiens*]."[7]

Such hypotheses are very interesting, though scientists do not know enough about our ancestors and the factors that might have contributed to the increase of brain size. However, more recently, Cornélio et al. have argued that, contrary to the hypotheses of Herculano-Houzel and Fonseca-Azevedo and Aiello and Wheeler, cooking food or the incorporation of greater amounts of animal products were not the main contributing factors to the human increased brain size. These researchers rely on neuroanatomical data as well as archeological evidence "to show that the expansion of the brain volume in the hominin lineage is described by a linear function independent of evidence of fire control, and therefore, thermal processing of food does not account for this phenomenon."[8] They show that human brain increased encephalization during evolution in fact occurred millions of years *prior* to the time when humans began cooking food. They argue that cooking food is unlikely the primary catalyst of the increase in human brain size. In their own words, these researchers point out that, "early hominins are likely to have obtained enough energy to sustain a large brain on a raw-food diet with 5–6 h of foraging per day."[9] Cornélio et al. also point out that their current data refute previous data that supporting the hypothesis that cooked food offered more energetic gain than raw food. They cite a particular 2011 paper by Carmody et al.[10], who suggest that cooking tubers and meat increases energetic gain in mice. Cornélio et al. "repeated the same experiment with meat and failed to find such energetic benefit in the thermal processing of meat." Furthermore, they point out that in a 2009 paper, Carmody and Wrangham indicate that such studies are scarce and often contradictory.[11]

Thus, Cornélio et al. suggest that rather than cooking food, it is more likely that brain expansion was favored by longer and longer hours of foraging per day and by the rise in the use of tools. Presumably, longer hours of foraging suggests that early humans had to walk farther and farther to find food and in so doing develop memory and other such cognitive skills required to improve foraging efficiency. Another suggestion that the researchers make is that another factor that accounts for foraging efficiency is a system of cooperation among early humans. Yet another important factor that contributed to the expansion in brain size in my view, which is corroborated by scientific evidence, is certain physiological mechanisms that improved nutrients absorption. Referring to the 2011 research of Fedrigo et al.[12] they suggest that, "For instance, differential expression of glucose transporters in the human brain could facilitate energy allocation."[13] In sum, the current scientific literature shows that the hypotheses that cooking food (especially of animal source) as the sole catalyst factors that contributed to brain expansion during evolution are not supported by archeological or physiological evidence.

There are some very important observations to be made here. First, the study by Cornélio et al. states, "early hominins are likely to have obtained enough energy to sustain a large brain on a raw-food diet."[14] By "a raw food diet" they do not suggest a raw vegan diet. In fact, they state the following,

> Early hominins likely increased their foraging efficiency by varying their diets, including seeds and meat, which are more caloric foods than wild plants and fruits. We provide compelling evidence indicating that thermal food processing is unlikely to explain increases in the foraging efficiency of early hominins.[15]

Note an interesting aspect of this statement. The authors offer evidence to show that cooking meat was not the factor that promoted brain growth; in fact, they make clear that brain growth was likely due to other factors, such as the rise of tools, cooperation, and longer foraging hours, which increased foraging efficiency. And their increased foraging efficiency was promoted by their consumption of high-caloric food, including meat and seeds. Now there are at least three problems with the notion that raw meat provided early hominins a good level of energy to enable them to improve their foraging efficiency. One is that early hominins, arguably, did not have access to considerable and continual amounts of meat, especially prior to the development of reliable hunting skills. I would assume that meat was an occasional source of calories rather than a dietary staple. Second, consider the following three factors: a) the significant expenditure of energy required hunting; b) the difficulties and risks involved in hunting; c) the brain requires a considerable supply of glucose. Consequently, it would seem more plausible to assume that the main contributing food that favored the expansion of the human brain was fruit, which is high in glucose and easier to come by than meat, especially if it is considered that early humans lived in tropical areas where fruit abounded. Third, raw flesh is not easy to chew and even less easy to digest (not to mention that the taste of fruit—even fruit like durian—is way more pleasant than raw flesh, and especially the flesh of ancient wild animals was tougher than the flesh of domesticated animals). Consider the expenditure of calories required to hunt, to slaughter, and to chew and digest raw meat. Even if that were the case, it is hard to think how this might have satisfied the brain's demand for glucose. As Katherine Zink and Daniel Lieberman point out, the masticatory features of early humans (as well as modern humans, I want to say) were inadequate for chewing raw meat: "These derived masticatory features suggest that the genus *Homo* consumed foods that were easier to eat, requiring fewer, less forceful chews and reducing the need for high maximum bite forces."[16] Leslie C. Aiello and Peter Wheeler make the same observation: "Gut size is highly correlated with diet, and relative small guts [like human guts] are compatible only with high-quality, easy-to-digest food."[17]

What follows from all this? Let's review a few points:

1   Current research provides compelling evidence indicating that cooking in general, and the preparation of foods from animal origins in particular, were not the catalyst to the expansion of the human brain during the past 2 million years.
2   Prior to the development of reliable hunting skills, early hominins would not (presumably) have a steady supply of meat; and prior to fire control, they would not cook it.
3   Hunting requires a higher expenditure of calories than foraging where fruit is plenty and available.
4   Raw meat is hard to chew and difficult to digest. In short, our small guts are not compatible with meat, raw or cooked. Humans have relative small guts; and small guts are compatible with easy-to-digest food.
5   The brain of early hominins grew as a result of eating high-quality, glucose-rich food.

What evidently follows from these premises is that early humans' increased brain size was promoted principally by raw food; such food must be easy to come by, easy to digest, and high in glucose. If we consider that raw meat is not easy to come by and does not satisfy the brain's high demand for glucose, it follows that fruit caused encephalization, and therefore it is the most efficient and optimal food for humans—the food to which humans are adapted.

I predict resistance here. It may be disputed that fruit is the *only* factor that promoted encephalization and that it is humans' optimal food. In addition, one may also feel that my conclusion seems too hasty. First, I would like to point out that fruit was not the *only* factor that promoted encephalization. As I mentioned previously, during evolution, humans developed certain physiological mechanisms that improved nutrients absorption, especially glucose transporters. Also, there is the contribution of foraging and toolmaking. My point is that the studies discussed earlier on present compelling evidence showing that, to use the title of a key study, "Human Brain Expansion During Evolution Is Independent of Fire Control and Cooking." Second, my argument does not commit the *appeal to nature fallacy*. That is, it does not move from the premise that fruit is natural to the conclusion that it must be morally acceptable or that we ought to eat fruit. Rather, to support my conclusion, I present a cumulative case:

1. There is compelling evidence that human encephalization occurred prior to—and independently of—cooking in general and the preparation of foods from animal origins in particular. And since humans have not undergone significant biological change in the last 40,000 or 50,000 years. It follows that humans are adapted to thrive on raw diets. The importance of raw diets, especially the role of fruit in providing optimal nutrition, is so important that cannot be dismissed by the alleged charge of an appeal to nature.
2. The scientific literature clearly shows that raw diets, especially plant-based raw diets, provide optimal nutrition and are beneficial in the prevention of diseases. On the other hand, it has become evident that animal-based food is not necessarily conducive to good health and in many case it is deleterious to human health.
3. Animal-based food produces negative aesthetic externalities, including violence and environmental degradation. Since these are not desirable, we should avoid animal-based food whenever and wherever possible.

Therefore, these three arguments combined make a strong case for plant-based raw diets.

Now, what further evidence do we have that fruit was responsible for the enlargement of the human brain? A new study has recently emerged to corroborate this conclusion. Alex DeCasien originally set out to test whether monogamous primates had bigger brains than more promiscuous species.[18] DeCasien collected data for the diets and social lives of more than 140 species of monkeys, apes, lorises, and lemurs. She asked which features were more likely to be correlated with bigger brains. Monogamy, promiscuity, group size, and other factors did not seem to predict anything about a primate's brain size. The one factor that predicts which species have larger brains is whether the subjects consumed primarily leaves or fruit: "what a species eats predicts its brain size—specifically that species whose diet is predominantly made up of fruit have, on average, larger brains than those that specialize on eating leaves."[19]

While many scientists find the news surprising, I don't find it surprising at all. It is logical to assume that in order to have a bigger brain, early hominins had to have a change in diet. Tubers and other starchy plants and raw meat are hard if not impossible to digest. Breaking down those foods takes a lot of time and energy. On the other hand, eating fruit is the quickest and most efficient way to acquire calories, that is, fruit is plentiful and easy to come by in tropical areas, it is easy to eat, it is more satisfying to eat than a raw piece of flesh or a starchy plant or tuber, and it is very easy to digest. And considering that the brains primary fuel is glucose, fruit benefitted early humans by offering a reliable and easy-to-come-by source of energy for the brain, which is what contributed to its growth. Also, I want to suggest that there is further evidence to support the thesis that fruit is the human diet. I will mention a few examples:

1 Najjar et al. 2018: A defined, plant-based diet utilized in an outpatient cardiovascular clinic effectively treats hypercholesterolemia and hypertension and reduces medications. This study shows the result of the implementation of a raw vegan diet for four weeks in a clinical setting with 30 patients. Granted, four weeks may be too short a time and 30 people too small a sample size. At any rate, the researchers found that despite such a short period of time, patients were greatly benefitted by a raw vegan diet. They concluded,

   A defined plant-based diet can be used as an effective therapeutic approach in the clinical setting in the treatment of HTN, hypercholesterolemia, and other cardiovascular risk factors while simultaneously reducing overall medication usage. Patients may find this therapeutic approach preferable to conventional and costly drug therapy. Further replication trials are needed with larger sample sizes, control groups, and other dietary comparison groups.[20]

2 He et al. 2007: Increased consumption of fruit and vegetables is related to a reduced risk of coronary heart disease: meta-analysis of cohort studies. This study also speaks for itself. The researchers found what I think is obvious:

   Increased fruit and vegetable intake in the range commonly consumed is associated with a reduced risk of stroke. Our results provide strong support for the recommendations to consume more than five servings of fruit and vegetables per day, which is likely to cause a major reduction in strokes.[21]

3 He et al. 2007: Increased consumption of fruit and vegetables is related to a reduced risk of coronary heart disease: meta-analysis of cohort studies.

   In conclusion, our meta-analysis of prospective cohort studies demonstrates that increased consumption of fruit and vegetables from less than 3 to more than 5 servings/day is related to a 17% reduction in CHD risk [coronary heart disease], whereas increased intake to 3–5 servings/day is associated with a smaller and borderline significant reduction in CHD risk. The average fruit and vegetable intake in most developed countries is approximately 3 servings/day, and it is even less in developing countries. The current recommendations are to increase the intake to 5 or more servings/day. Our results provide strong support for these recommendations. If these were achieved, there would be a large reduction in CHD morbidity and mortality. As raised blood pressure throughout its range is a major cause of CHD, it is likely that the blood pressure-lowering effect of fruit and vegetables is the major mechanism that contributes to a reduced risk of CHD. In addition to its effect on CHD, an increase in the consumption of fruit and vegetables may reduce the risk of strokes and some cancers and may have other health benefits, for example improving bowel function, helping people adhere to weight-reducing diets through improved satiety.[22]

The point of the foregoing is to show that consumption of fruit results exactly in what one would expect to see; that is, fruit provides high-quality calories in an easy-to-digest package, is conducive to good health. Virtually all studies that involve an increase in the consumption of fruit show positive results in terms of human health. On the other hand, it is interesting to note that while there exist studies that show possible benefits from eating certain animal products, most studies show that consumption of animal products is not exactly beneficial to our health. In fact, more and more dieticians and physicians nowadays are warning their patties to consume animal products in moderation.

Consequently, there are no good reasons to believe that it is the raw meat that contributed to the expansion of human brain and there is compelling evidence that consumption of raw meat is in no way beneficial to the human body, nor is raw meat essential to maintain high cognitive capacities. Therefore, the foregoing discussion offers significant evidence that fruit in particular, and to some extent tender leafy greens, constitute the human diet.

My point here is that, since there is an optimal diet for every species of animals on earth, and humans are just animals, it follows that there is a specific diet for humans. In my view, such a diet is what I call the "human diet," which is a raw vegan diet. I call it human diet instead of raw vegan because I want to emphasize a philosophical (some might say a scientific) point, that is, nutrition is a physiological requirement based on the specific biological needs of an organism. Raw foodism and veganism focus on culinary and psychological needs of people. I do not intend to undermine the importance of gustatory and psychological needs. Unfortunately, they often fail to align with our biological needs, and lead to many health problems. Moreover, our psycho-culinary needs often lead to immoral practices in the name of taste, such as the exploitation of animals.

So, what is the human diet? I have to explain three points: 1) what "optimal diet" and "specific" mean; 2) what the human diet is; and 3) the basis for asserting that the human diet is optimal and specific for humans.

1 Optimal and specific diet: A few years ago, one spring day I was strolling through Central Park here in New York City and noticed a man sitting on a park bench speaking on his cellular phone while holding a container of French fries with the other hand. Then I noticed that a squirrel had crawled up on the bench to take advantage of the man's being distracted by his phone conversation and eat his fries. Squirrels can eat fries, peanuts, and other human junk food. However, neither of those foods is optimal for squirrel health.[23] Not surprisingly, squirrels thrive when they eat food that is optimal for them. By optimal I mean that it is not harmful to their health and they could eat in abundance without experiencing negative health repercussions. In other words, a food is optimal if it is the food that an organism adapted to eat. People nowadays, just like hominins millions of years ago did, eat and digest

many food items, but it does not mean that those foods are optimal. To be optimal a food for a species, a food must contain valuable nutrition, be non-toxic to that species, and be consumed in its whole state. A food that is optimal, thus, is said to be specific for a species.

2   The human diet: What makes a food specific is that that particular species evolved to digest and benefit from a particular food. Squirrels evolved in such a way that mostly nuts and fruits are specific for them. The human diet is a diet that human beings evolved to benefit from, a diet of fruit, especially sweet fruit, and tender leaves. By fruit I refer to the sweet or savory seed-bearing products of trees and plants, such as avocados, mangoes, tomatoes, peppers. Naturally the most beneficial fruits are those that are edible in their raw state and easily digestible. For example, botanically, an eggplant is a fruit, but in its raw state is not easily digestible. Occasionally, humans can benefit from small and sporadic amounts of unprocessed nuts and seeds. The human diet does not include anything that is toxic or useless to the human body, such as salt, alcohol, spices, caffeine, and food that is processed. Once again, these foods and more might be psychologically important, but nonetheless they are harmful and useless to the body. Arguably the human diet strikes one as boring or lacking in variety. Two important points, however, must be borne in mind: a) all animals in nature eat a rather unvarying diet consisting in very few foods; and b) what is important is not so much variety per se, but rather it is important that an organism acquires all nutritional needs from a diet. The human diet is specific and optimal for humans because fruits and tender leafy greens are easily digestible and contain all the important nutrients.

3   The basis for asserting that the human diet is optimal and specific for humans: On what grounds, then, is the human diet an optimal and specific diet for humans? I propose two empirical grounds. The first one is somewhat speculative because it has to do with a rational consideration of *Homo sapiens'* dietary needs based on human physiology and anatomy. The second is a consideration of scientific data showing that consuming raw fruit and tender leafy greens is conducive to good health. This however is incomplete due to the lack of extensive research. However, the available research shows exactly what we expect to see—that fruit is consistently beneficial to human diet.

What further evidence is there to show that human beings' ideal diet is the human diet? Consider the following. Scientists have estimated that there are approximately nine million species on earth. Despite their being different, they share common aspects about diet. In my view at least, if nine million different species have something in common, this fact merits attention. I am referring to three facts: 1) all animals have a specific diet, 2) no animals cook food, 3) their respective diets do not vary much. From these three common aspects, the second step of my argument is that human beings, just like all other animals, are the product of their natural environment. What I mean is that humans are very much part of nature, despite the fact that modernity and technology has transformed humans and alienated them from nature.

The next step is to consider that human physiology has not undergone significant evolutionary changes. Some for example suggest that long distance running has contributed to increases in red blood cells and that over time many human beings became capable of tolerating and digesting starchy foods.[24] Another example is that humans developed osteoarthritis and degenerative arthritic issues of the spine and lower limbs as a result of upright posture.[25] However, human beings have not gone through significant changes that might have equipped to digest cooked animal- and plant-based food. As the famous American paleontologist and evolutionary biologist, Stephen Jay Gould pointed out, "There's been no biological change in humans in 40,000 or 50,000 years. Everything we call culture and civilization we've built with the same body and brain."[26] In other words, while humans today are more technologically sophisticated than their ancestors, they have similar brains. Moreover, while humans today cook their food, for millions of years prior to cooking humans had been eating a raw diet. The fact however is that cooking food is a relatively new practice for humans and the human body has not had the time to undergo any significant physiological changes that might enable them to process and benefit cooked food. Consequently, there is no reason to believe that cooking represents an optimal or specific diet for humans. Thus, the human diet is, as I describe it, the optimal and specific diet, a diet consisting in uncooked fruits and tender leafy greens.

Humans are indubitably frugivores. We evolved from the same genetic line as our closest relatives, the bonobos and chimpanzees, whose diet consists in fruit and leafy greens. Our digestive anatomy is very similar to theirs. Granted, chimpanzees are also known to eat the flesh from other animals, for example monkeys. However, their primary food is fruit and tender leaves. As I already mentioned, humans are capable of digesting cooked food, but this doesn't change the fact that our digestive anatomy is designed to eat fruit.[27] Apes are also capable of eating some cooked food. They seem to prefer it when offered, but no primatologist would suggest that cooked food is optimal for apes.[28] Why should it be optimal for humans, then? Most researchers agree that the few thousands of years during which our species spent as foragers had no significant effects on our digestive anatomy.[29] In other words, modern humans eat cooked food, but their digestive systems are still the same system that nature designed for eating fruit. As primatologist Katharine Milton writes, "The widespread prevalence of diet-related health problems, particularly in highly industrialized nations, suggests that many humans are not eating in a manner compatible with their biology."[30] It is not surprising, then, that modern humans are afflicted by numerous maladies, which are caused by humans' deviation from "a diet on which the primate line has flourished for many tens of millions of years and *produced them*"[31]

Cooking enables the consumption of foods that are inherently indigestible or toxic. However, cooking is in no way beneficial to humans. For example, enzymes are important for proper digestion. Unfortunately, they are heat sensitive, which means that cooking at temperatures over 117°F (47°C) deactivates enzymes.[32] Cooking also deactivates and destroys water-soluble vitamins, such as vitamin C, B;

minerals and vitamin A are also lost during cooking.[33] Phytonutrients are natural chemical compounds contained in plants; they are what give fruits and vegetables their characteristic color. These chemicals protect the body and fight diseases.[34] Phytonutrients in freshly harvested plant foods can be destroyed or removed by cooking.[35] Cooking food, especially starchy food, can cause the formation of acrylamides, which are substances that have the potential to cause cancer.[36] Furthermore, there is a growing body of science indicating that raw food diets have the potential to reduce the risk for cardiovascular disease, can help turn off the gene Chromosome 9p21, which is the most potent genetic associations with heart disease, improve mental and emotional quality of life. On the other hand, cooked foods, which are by definition high in proteins, tend to shorten one's life.[37]

Cooking denatures protein. Denaturation is a modification of the molecular structure of protein by heat or by an acid that destroys or diminishes its original properties and biological activity. Denaturation means that the molecular structure of proteins is modified and, as a result, the modified structure can be harmful for the body. Thus, "Most genetic diseases can be linked back to a protein that does not have the structure it should."[38] When food is cooked, resistant linkages are formed between the amino acid chains that the body cannot separate.[39] Also, the Maillard reaction negatively affects the food (typically starches) generating prooxidants, carcinogens, and lowering the nutritional value of food.[40] Furthermore, cooked fats can be rendered rancid and carcinogenic.[41]

Thus, while humans have been cooking food for a long time, cooked food is by no means optimal for our species. To be sure, I want to emphasize the fact that the research on this topic is gargantuan. One can easily find many studies that show the benefit of cooked food. I did not include such studies in my discussion, but not because I feel that they are a threat to my argument that cooking food is contrary to our nature. Rather, people are informed about the benefits of cooking food every day because most, if not all, societies in the world cook food. Such studies show that cooking some food increases the bioavailability of certain nutrients, etc. However, it is typical to ignore the side effects of cooking food, some of which I discussed in the previous section. The important point is that virtually all studies on raw food confirm that eating fruits and leafy greens in their raw state is highly beneficial to our health. There are at least four objections to raw veganism:

1 Many object to the notion that cooking destroys nutrients. For example, they point out that cooking tomatoes increases by many times the bioavailability of an antioxidant called lycopene. This objection, however, is vacuous. For, it assumes that more is synonymous with better. This is not necessarily true. For example, researchers show that "High-protein diets may be appropriate for some individuals, but not for others; hence, specific individual needs, as well as potential negative consequences, must be considered cautiously before such a diet is adopted."[42] By the same token, all nutritional requirements can be met on a raw diet, including lycopene.

Secondly, a careful study of the research clearly shows that there is no doubt that cooking food denatures or destroys nutrients.

2   A study suggests the possibility of low bone mass in long-term raw vegans due to vitamin D deficiency.[43] First, consider two problems: first, there is practically no research on long-term raw veganism; second, people label themselves raw vegans when in reality they are not raw vegans. They eat cooked food, and often they occasionally even consume animal products. Thus, it is hard to imagine how the study can possibly be informative. Second, upon review of pertinent literature, it is evident that there is "no association between s25(OH)D concentrations and vegetarian status in either our black or white cohorts. This would indicate that factors other than diet have a greater effect on s25(OH)D than vegetarian status."[44]

3   Another concern is that a raw vegan diet is limited in the variety of foods. For example, raw foodists do not consume beans, legumes, whole grains, and potatoes. This just begs the question. For one thing, prior to cooking food human beings ate a raw diet. Granted they might have had the occasional dead bug. But in a tropical setting where fruit abound, what else could humans eat if not fruit? For millions of years human beings acquired all the required nutrition from plants. Furthermore, science shows that fruits and vegetables provide all the essential nutrients. Second, while cooked starches and legumes contain valuable nutrients, at the same time, as I showed earlier, they can be harmful in many ways.

Some argue that raw vegan diets can be harmful as a result of nutrient deficiencies and often under-eating.[45] Granted, under-eating leads to deficiency. But this is hardly news. The researchers also conclude that a raw food diet leads to high loss of body weight, and therefore it can't be recommended on a long-term basis. I have a two-pronged reply: first, since many people in the world are overweight or obese, the fact that a raw vegan diet leads to weight loss cannot be but a virtue of such a diet; second, some people believe that weight loss or gain is a function of some mysterious principle of the universe. In reality, each individual requires a certain number of calories to maintain his or her weight; the number of calories varies depending on many factors, including one's lifestyle. Thus, consuming more calories than one requires will lead to weight gain; by the same token, consuming fewer calories will lead to weight loss. But it is not a raw vegan diet that magically leads to weight loss. Rather, it is the number of calories that one eats.

To conclude the foregoing discussion, consider this further point. In 1930, Dr. Paul Kouchakoff discovered an increase of leukocytes, white blood cells, in subjects that ate cooked food, while subjects that ate raw food had no change in white blood cells count.[46] This phenomenon is known as digestive leukocytosis. Leukocytosis typically occurs as a result of an inflammatory response. It is interesting that it also occurs as a result of consuming cooked food. What is the explanation? Most agree that the most plausible and simplest explanation is that cooked food is unfit for human consumption.[47] In other words, cooked food is not the human diet.

Raw and veganism

Raw vegan diets have been around since the early 1900s, though they have become more and more popular in the past couple of decades. In this era of dietary confusion, people have tried to figure out a diet that is really conducive to health benefits. Since most food nowadays is processed, and unhealthful, the logical conclusion is to eat as much whole food as possible. This led many to adopt raw veganism. Predictably, medical and scientific authorities have shown resistance to raw veganism and warned the public about the potential danger of eating a raw diet. This should not come as a surprise considering that virtually all societies in the world revolve around cooked food. In an article titled "Top 5 worst celeb diets to avoid in 2018", the British Dietetic Association lists five raw diets apparently followed by many celebrities and argue that such diets could compromise long-term health.[48] I do not deny that some vegan diets can be wacky, but I want to make three remarks about evaluating raw diets because there is a great deal of misconception about raw veganism.

First remark

Raw veganism is just an umbrella term. There are many diets that are considered raw. For example, some people known as fruitarians consume a diet of, well, fruit. Some eat only fruit, while others include greens; some others include nuts, seeds, fermented, and dehydrated foods, such as dry fruit, dry kale, dry mushrooms, and more, while others avoid them. Yet others eat predominantly greens and limit fruits, especially very sweet fruits. Some eat a very high-fat diet, including nuts, seeds, and avocados, durian, and others eat a very low-fat diet excluding those items. Some people consume herbs and spices, arguing that these are beneficial to our health, others arguing the opposite stay away from herbs and spices (most spices are not raw). Also, some individuals who follow raw vegan diets may include honey, though honey is not a vegan food. In fact, some raw vegans consume shellfish, such as mussels and clams, arguing that they are not sentient organisms. Furthermore, some label themselves as "raw vegans" despite consuming limited amounts of cooked food and even raw or cooked animal-based food. In short, people tend to make up their own definition of raw veganism making it a complicated matter to determine whether or not a raw vegan diet is conducive to good health.

Second remark

There are many food items that are raw, but are not necessarily healthful. Since research is elusive about the nutritional benefits of these foods, I will refrain from cherry-picking scientific studies. In my view, these are not necessarily harmful foods if consumed occasionally; at the same time, they are not healthful foods, and should not be a staple in any healthful diets. These foods can be grouped under at least three categories: The first category is fermented foods.

Raw foodists may consume fermented foods and beverages, such as krauts, kombucha, and other pickled products. These products may be slightly alcoholic, and contain various amounts of sodium chloride (table salt). Our bodies need sodium but not sodium chloride. Sodium chloride is a mineral—a rock. Our bodies can be seriously damaged by table salt. Sodium, on the other hand, is a mineral that is naturally occurring in plants. Plants can absorb rock minerals more efficiently than humans, but the human body can absorb only minerals from organic sources. Thus, the ideal situation is to consume sodium by eating organic sources, i.e., celery, coconut water, tomatoes, melons, leafy greens, etc. The bottom line is that salt was introduced as a preservative because salt kills bacteria, especially during times when refrigerators did not exist, to prevent meats from rotting. In practical terms, salt is not a food.

The second category includes products labeled "raw" that are not raw or are exposed to heat and dehydrated. Since 2007 most nuts and seeds in the USA must now be steam pasteurized before sale, though the law allows companies to label their nuts and seeds as raw. Nuts can be healthful if consumed sparingly. However, I have some concerns about consuming nuts. First, as will be explained later, cooking changes the molecular structure of food in ways that are often more damaging than beneficial. Cooking nuts can cause oxidation, acrylamides, which can be harmful for the body.[49] Second, nuts are calorically dense, high in protein and fats. While in certain circumstances such characteristics are desired, ideally one should follow a diet that is low in calories and protein.[50] Third, nuts are nearly devoid of water content, and this may lead to dehydration. Ideally, a diet should be low in calories and proteins and high in minerals, vitamins, antioxidants, and water; thus, fruit is a more healthful option than nuts in most cases.[51]

Another food that is marketed as raw is cacao. Cacao beans are fermented and roasted to achieve the characteristic flavor. Once again, in my view cooked nuts, seeds, and cocoa are not conducive to good health. I concede that they can be healthful in their raw state and in limited amounts, except for cocoa because it contains the stimulant alkaloid theobromine. The point here, however, is that the nuts and seeds that many raw vegans consume are not raw. The second group of raw foods is dehydrated products. These foods are prepared by exposing them to dry heat; this is typically achieved by processing fresh fruit, vegetables, legumes, and more, in dehydrator devices. Many argue that heated food can still be considered raw, and the nutrients are not compromised, so long as the temperature never goes above $115°F$. There is no legal regulation, or agreement among raw vegans, about the maximum permissible temperature for a food to be considered raw. Some say the maximum is $115°$, but others say that $118°$ or even $120°F$ will not destroy micronutrients. However, vitamins like vitamin A, C, and thiamin are very sensitive to air and heat and can be significantly damaged or destroyed.

The third category is stimulants and drugs. Stimulants can increase alertness or give the feeling of increased energy. Take coffee for example. One eight-fluid-ounce cup of coffee contains only one calorie. In other words, coffee confers to the body no energy. The reason one feels more alert and energetic is that coffee stimulates the adrenal glands, which leads to cortisol production, and in turn it leads to stress and adrenal fatigue. Not to mention that coffee is not raw since it is a beverage obtained from ground-roasted seeds boiled in water. Tea is also a stimulant and contains caffeine. In addition to coffee and tea, some vegans consume alcoholic beverages. I don't believe (I hope) that I have to cite any kind of scientific study here to show that alcohol is not food. Alcohol, even in moderation, is poison and it kills everything that it gets in contact with, and that's why it is used as a disinfectant.[52] Some people may advocate alcohol in moderation. However, it is advocating taking poison in moderation.

Furthermore, raw vegans may take medications, use drugs, sleep too little, breathe toxic fumes, and so on. The bottom line is that drugs, alcohol, coffee, etc., are not foods. Also, some raw vegans may not sleep enough hours, not be exposed to the sun long enough, be stressed, eat too much salt, etc. When dietetic associations or other medical and scientific entities assess the impacts of raw vegan diets on human health, they often fail to account for a holistic picture of the individual's diet. What I mean is the following. An individual's diet can reveal a great deal about his health. However, his diet alone cannot give a full picture of his health. For example, there are many cases of people who are unhealthy while following very healthful diets. In many cases the culprit may not be the diet itself, but rather one or more of the factors just mentioned. Typically, raw vegans tend to be concerned about their health; still, some of them drink alcoholic beverages, smoke tobacco, or take drugs. The bottom line is that there may be numerous factors that make raw vegan diets appear unhealthful.

Third remark

Research on the potential benefits or drawbacks of a raw vegan diet is lacking or is biased or is confusing. What I mean by "biased" is that non-vegans who eat cooked food and use standard cooked diets are used as a point of reference to conduct the research. Also, in most cases the researchers use questionable sample populations that may or may not be strictly vegans or, as I mentioned above, may consume drugs, stimulants, and other unhealthful substances and foods. It is true that there is a considerable body of research that vegan diets can achieve weight loss, reverse diabetes, and lower cholesterol. However, at the same time, the so-called paleo diets and even meat-and-dairy diets seem to lead to good health. Ultimately, in my view at least, it is vague to say that this or that diet is beneficial. What seems to be beneficial without a doubt or need of scientific corroboration is consuming lots of fresh, uncooked fruit and tender leafy greens. At any rate, what scientific research establishes is a complicated matter.

Essentially, there exist studies validating the efficacy and benefits of either dietary approach. Having read hundreds of studies on diet and nutrition, it is experience that reveals the true flaws or virtues of such studies. One of the main problems with nutrition research is that most of the studies rely on the dietary patterns of people's statement obtained through questionnaires. Obviously, it is hard to monitor large populations and ensure that they are following certain dietary protocols. So, researchers must trust what the individuals surveyed tell them. As I mentioned, in addition to diet there are factors that have an enormous impact upon our health, such as stress, drugs, stimulants, lifestyle, etc. Thus it is not easy to know what causes certain conditions. These types of population studies also cannot prove cause and effect, but merely show correlation. Furthermore, laypeople and the media distort the science and create myths that serve their respective interests. The farm animals' sector has an interest in selling animal products; and people who consume animal products have an interest in justifying eating animal products. Nowadays, we hear that we have to consume calorie-dense foods high in fats and proteins to be in good health, when in fact thinking logically reveals that the opposite is true. Many health magazines inspired by certain scientific studies promote dietary cholesterol and recommend consuming eggs, dairy products, and meats. Is this a coincidence? The meat industry needs to sell more animal products, people are told to eat more animal products, and people buy more animal products. Why are we never told to eat more blueberries, for example? Naturally, one may point out that my observations about scientific studies go both ways. That is to say, studies that find plant-based diets beneficial for our health suffer from the same problems just mentioned in connection with studies in favor of animal-based diets. The only way to make sense of scientific studies about nutrition is through experience. Furthermore, a peculiar aspect of this issue, which speaks in favor of plant-based diets, is the fact that while there is evidence that animal products are not beneficial for human health, there are no studies that show that fruit and vegetables can be deleterious for human health.

Conclusion

In this chapter I showed that the mainstream view that early hominins became human beings with an increased brain size as a result of cooking food and the incorporation of animal products in their diets is false. I proposed compelling evidence and argumentation to show that early hominins benefitted from a raw diet of mostly fruit, which promoted improved foraging techniques and, in their turn, cognitive development. Two implications follow. The first one, which I forcefully argued, is that the ongoing question of which is the optimal diet for humans can be answered confidently, "Fruit!" The second implication, which I merely suggest, is that perhaps my discussion provides the basis for a discussion of the human diet with respect to our treatment of non-human animals and the environment. In other words, if human beings are designed, as it were, to eat fruit, perhaps this fact gives us an evolutionary reason to justify ethical raw veganism and condemn the exploitation of non-human animals and the environment.

What's next: we have to save veganism

We have to save veganism from becoming a commercial food trend. Veganism today has become a giant meatball (pun intended) made of many ingredients, including pop philosophy and ethics, pseudo- science, lucrative business, foodism, and fashion. In other words, veganism has shifted its focus from the animals to people's self-indulgence. By definition, I am a vegan because I do not consume animal products; but I often do not identify myself as a vegan in order to avoid possible misinterpretations. Moreover, I am critical of certain aspects of the current state of veganism. Many people are vegans on the ground of vague moral principles or follow veganism under the assumption that a vegan diet is automatically healthful or even because they simply like vegan food. In the next chapter, I will discuss two concerns about the direction of veganism: 1) veganism can promote consumerism; it promotes intemperance and self-indulgence, and ultimately it does not help the environment. One way I propose to reestablish the true spirit of ethical veganism is by embracing veganarchism, which, in my view, is the combination of virtue ethics and anarchist values; 2) I suggest a valiant solution to the environmental problems that the meat industry causes—the solution is abolition of the production of animal products through a virtue-oriented education.

As a concluding remark, I want to give a quick overview of my concern about the direction of veganism. More and more people become vegans or consume vegan products. Should vegans be happy about it? Perhaps not so fast! As the number of vegans is on the rise, many non-vegan corporations are jumping on the vegan bandwagon by creating new foods tailored to the vegan community. Just this year, MacDonald's has developed a vegetarian "Happy Meal", TGI Friday's has announced a "bleeding" vegan burger, and Gregg released its controversial vegan sausage roll. Also, the companies Impossible and Beyond Meat have been producing vegan burger patties, meatballs, and vegan ground beef, that are supposed to emulate the smell, taste, and texture of real meat. Impossible's products are not yet available for sale, but are served at many vegan joints and at many non-vegan fast-food chains, such as White Castle, Applebee's, Bareburger, 5 Napkin Burger, Burger King, to name a few. Beyond Meat's products are available for sale in many supermarkets. One of the oddities regarding such products is that they are produced to emulate real meat products. Beyond Meat burger patties are packaged in the same way as meat patties. Also, Beyond patties look uncannily like regular meat patties, including the blood aspect, which is achieved by the addition of beet juice. Beyond Meat burgers, when cooked, react to heat the same way as meat in that they change color from blood red to the characteristic brownish color of cooked meat that obtains as a result of the Maillard Reaction. A further peculiarity regarding the sale of Beyond Meat products is that they are often placed in the same refrigerators where meat products are kept. It is clear that the resemblance to meat products of many vegan products and the marketing strategies that these companies adopt are counterproductive for veganism.

Arguably the existence of these products does two things: one, it sends a specific message to the public—this message is that meat is the real deal, and then there are products that emulate meat, reinforcing the notion that eating animals is the norm, while veganism is an "option." Vegan products are viewed not as the real deal, but a second-tier product or a second best or a weird brother that gives meat eaters a wink. In fact, most vegan foods emulate animal products—a trend that renders some vegans an odd bunch of individuals who refuse to eat animal products but wish to eat animal products and in order to do so they eat foods that taste, look, and smell like animal products.

Regardless of whether these products are tasty or popular or are helpful as transitioning foods, the main problem is that veganism is no longer the moral lifestyle choice of eco-militants, but rather another trend that has successfully entered the mainstream. Instead of representing the opposition, veganism runs the risk of becoming an ally to the meat industry. Veganism revolves so much around food that, ironically, it has been swallowed and digested by mainstream non-vegan corporations. One might note that vegan products, such as burgers, vegan cheese that melts and stretches, and many others, have been produced to please meat eaters more than vegans. As I noted, many of these products taste and look almost identical to their animal-based counterparts. I, as well as many others, became a vegan because I am repulsed by the smell and the idea of killing animals to cook and eat their bodies. At any rate, one thing is clear: these corporations are not ethical enterprises, but just corporations. They are concerned about profit– not about morality, not about the animals, and not about the environment. Unfortunately, the promotion of such products perpetuates carnist values, that is, the ideal of meat and animal products as the norm. Veganism has to be saved from being misappropriated by the meat industry. As an indication of the future of veganism, consider Arby's "Megetable"—a piece of meat that looks like a carrot![53]

All this sounds like a market ploy perpetrated by the meat industry to first eclipse the moral nature of veganism and then take control of the same. Furthermore, a very important point is that by purchasing a vegan meal from a large food chain, vegans support businesses that can continue further environmental destruction through the production of animal products. Clearly, instead of stopping animal-based companies, the purchasing of the vegan option in fact increase the profit of such companies, which will continue their environmentally damaging businesses.

Notes

1 E. Meyer-Renschhausen and A. Wirz, "Dietetics, health reform and social order: vegetarianism as a moral physiology. The example of Maximilian Bircher-Benner (1867–1939)." *Medical History*, 43(3), 1999: 323–341. doi:10.1017/s0025727300065388
2 K. M. Cummings and R. N. Proctor, "The changing public image of smoking in the united states: 1964–2014." *Cancer Epidemiology, Biomarkers & Prevention: A Publication of the American Association for Cancer Research*, 23(1), 2014: 32–36.

doi:10.1158/1055-9965.EPI-13-0798; J. Blocker et al. *Alcohol and Temperance in Modern History*, Vol. 1. Santa Barbara, CA: ABC-CLIO, 2003: xxxi-xiv.
3   M. Fitzgerald, *Diet Cults: The Surprising Fallacy at the Core of Nutrition Fads and Guide to Healthy Eating for the Rest of US*. Pegasus Books, 2014: 31–33.
4   Smithsonian National Museum of Natural History, "Introduction to Human Evolution." 2019. http://humanorigins.si.edu/education/introduction-human-evolution
5   G. Lawton, "Every human culture includes cooking – this is how it began." *New Scientist*, 2016. www.newscientist.com/article/mg23230980-600-what-was-the-first-cooked-meal/, para. 6.
6   K. Fonseca-Azevedo and S. Herculano-Houzel, "Metabolic constraint imposes tradeoff between body size and number of brain neurons in human evolution." *Proceedings of the National Academy of Sciences*, 109(45), 2012: 18571–18576. doi:10.1073/pnas.1206390109
7   Leslie C. Aiello and Peter Wheeler. "The expensive-tissue hypothesis: The brain and the digestive system in human and primate evolution." *Current Anthropology*, 36 (2), 1995: 199.
8   Cornélio, A. M., de Bittencourt-Navarrete, R. E., de Bittencourt Brum, R., Queiroz, C. M. and Costa, M. R. "Human brain expansion during evolution is independent of fire control and cooking." *Frontiers in Neuroscience*, 10, 2016: 167. doi:10.3389/fnins.2016.00167: 1.
9   Cornélio et al., 2016.
10 R. N. Carmody, G. S. Weintraub and R. W. Wrangham, "From the cover: Energetic consequences of thermal and nonthermal food processing." *Proc. Natl. Acad. Sci. U.S.A.*, 108, 2011: 19199–19203. doi:10.1073/pnas.1112128108
11 R. N. Carmody and R. W. Wrangham, "The energetic significance of cooking." *J. Hum. Evol.*, 57, 2009: 379–391. doi:10.1016/j.jhevol.2009.02.011
12 O. Fedrigo, A. D. Pfefferle, C. C. Babbitt, R. Haygood, C. E. Wall and G. A. Wray, "A potential role for glucose transporters in the evolution of human brain size." *Brain Behav. Evol.* 78, 2011: 315–326. doi:10.1159/000329852
13 Cornélio et al., 2016: 9.
14 Cornélio et al., 2016: 7.
15 Cornélio et al., 2016: 8.
16 Katherine D. Zink and Daniel E. Lieberman, "Impact of meat and Lower Palaeolithic food processing techniques on chewing in humans." *Nature* 531(24) March: 3.
17 Aiello and Wheeler, 1995: 199.
18 A. R. DeCasien, S. A. Williams and J. P. Higham, "Primate brain size is predicted by diet but not sociality." *Nat Ecol Evol*, 1, 2017: 0112.
19 DeCasien et al., 2017: 1.
20 Rami Najjar, Carolyn E. Moore and Baxter D. Montgomery, "A defined, plant-based diet utilized in an outpatient cardiovascular clinic effectively treats hypercholesterolemia and hypertension and reduces medications." *Clin. Cardiol.*, 41, 2018: 313.
21 F. J. He, C. A. Nowson, M. Lucas and G. A. MacGregor, "Increased consumption of fruit and vegetables is related to a reduced risk of coronary heart disease: meta-analysis of cohort studies." *J. Hum. Hypertens*, 21, 2007: 717–728.
22 He et al., 2007: 728.
23 "Tasty foods you can feed squirrels and what to avoid." *Feeding Nature*, 2014. https://feedingnature.com/tasty-foods-you-can-feed-squirrels-and-what-to-avoid/
24 Alexander Q. Vining and Charles L. Nunn. "Evolutionary change in physiological phenotypes along the human lineage." *Evolution, Medicine, and Public Health*, 2016 (1), 2016: 312–324. doi:10.1093/emph/eow026
25 Frank Rühli, Katherine van Schaik and Maciej Henneberg, "Evolutionary medicine: The ongoing evolution of human physiology and metabolism." *Physiology*, 31 (6), 2016: 392.

26 Stephen Jay Gould, "The spices of life: An interview with Stephen Jay Gould." *Leader to Leader*, 15, The Unofficial Stephen Jay Gould Archive, 2000: 14–19. www.stephenjaygould.org/library/gould_spice-of-life.pdf, 19.
27 My using the term "designed" is not meant in a religious or mystical sense, nor do I exclude those meanings. Rather, I merely acknowledge that humans and their digestive system are the products of evolution.
28 Katherine Milton, "Nutritional characteristics of wild primate foods: Do the natural diets of our closest living relatives have lessons for us?" *Nutrition*, 15(6), 1999: 488–498.
29 Katherine Milton, "Back to basics: Why foods of wild primates have relevance for modern human health." *Nutrition*, 16, 2000a: 481–483; Katherine Milton, "Hunter-gatherer diets: A different perspective." *American Journal of Clinical Nutrition*, 71, 2000b: 665–667.
30 Milton, 1999: 488.
31 Hans Georg Dehmelt, "What is the optimal anthropoid primate diet?" 2001. doi: arXiv:physics/0112009; Hans Georg Dehmelt, "Healthiest diet hypothesis." *Medical Hypotheses*, 64(4), 2005.
32 Nelson G. Chen et al. "Transient model of thermal deactivation of enzymes." *Biochimica et biophysica acta*, 1814(10), 2011: 1318–1324. doi:10.1016/j.bbapap.2011.06.010
33 D. Rumm-Kreuter, and I. Demmel, "Comparison Of Vitamin Losses In Vegetables Due To Various Cooking Methods." *J. Nutr. Sci. Vitaminol.*, 36, 1990: S7–S15; Jasraj K. Deol, and Kiran Bains, "Effect of household cooking methods on nutritional and anti nutritional factors in green cowpea (vigna unguiculata) pods." *Journal of Food Science and Technology*, 47(5), 2010: 579–581. doi:10.1007/s13197-010-0112-3
34 Charu Gupta and Prakash Dhan, 2014. "Phytonutrients as therapeutic agents." *Journal of Complementary and Integrative Medicine*, 11(3) 2014: 151–169. doi:10.1515/jcim-2013-0021
35 Jaqueline B. Marcus, Vitamin and mineral basics: The ABCs of healthy foods and beverages, including phytonutrients and functional foods: Healthy vitamin and mineral choices, roles and applications in nutrition, food science and the culinary arts. *Culinary Nutrition*, Academic Press, 2013: 279–331. doi:10.1016/B978-0-12-391882-6.00007-8
36 L.S. Jackson and F. Al-Taher, Effects of consumer food preparation on acrylamide formation. In Friedman, M. and Mottram, D. (eds) *Chemistry and Safety of Acrylamide in Food. Adv. Exp. Med. Biol.*, 561, 2005. Springer, Boston, MA.
37 C. Koebnick, A. L. Garcia, P. C. Dagnelie, C. Strassner, J. Lindemans, N. Katz, C. Leitzmann and I. Hoffmann, "Long-term consumption of a raw food diet is associated with favorable serum LDL cholesterol and triglycerides but also with elevated plasma homocysteine and low serum HDL cholesterol in humans." *J. Nutr.*, 135, 2005: 2372–2378.; Conor P. Kerley, "A review of plant-based diets to prevent and treat heart failure." *Cardiac Failure Review*, 4(1), 2018: 54–61. doi:10.15420/cfr.2018:1:1; Ron Do et al., "The effect of chromosome 9P21 variants on cardiovascular disease may be modified by dietary intake: Evidence from a case/control and a prospective study." *PLoS Medicine*, 8(10) 2011: e1001106. doi:10.1371/journal.pmed.1001106; K. L. Brookie, G. I. Best and T. S. Conner, "Intake of raw fruits and vegetables is associated with better mental health than intake of processed fruits and vegetables." *Front. Psychol.*, 9, 2018: 487; Sara Arganda et al., "Parsing the life-shortening effects of dietary protein: Effects of individual amino acids." *Proceedings. Biological Sciences*, 284(1846), 2017: 20162052. doi:10.1098/rspb.2016.2052
38 I. Rehman and S. Botelho, Biochemistry, secondary protein structure. In *StatPearls*. Treasure Island, FL: StatPearls Publishing, 2019. www.ncbi.nlm.nih.gov/books/NBK470235/

39 B. Alberts, A. Johnson, J. Lewis et al. *Molecular Biology of the Cell*, 4th edition. New York: Garland Science; 2002. Available from: www.ncbi.nlm.nih.gov/books/NBK21054/
40 Nahid Tamanna and Niaz Mahmood, "Food processing and Maillard Reaction products: Effect on human health and nutrition." *International J. Food. Sci.*, 2015, Article ID 526762. doi:10.1155/2015/526762
41 Samantha A. Vieira et al. "Challenges of utilizing healthy fats in foods." *Advances in Nutrition*, 6(3) 2015: 309S-17S. doi:10.3945/an.114.006965
42 Marta Cuenca-Sánchez et al. "Controversies surrounding high-protein diet intake: satiating effect and kidney and bone health." *Advances in Nutrition*, 6(3), 2015: 260–266. doi:10.3945/an.114.007716: 265.
43 L. Fontana, J. L. Shew, J. O. Holloszy and D. T. Villareal, "Low bone mass in subjects on a long-term raw vegetarian diet." *Arch Intern Med.*, 165(6), 2005: 684–689. doi:10.1001/archinte.165.6.684
44 J. Chan, K. Jaceldo-Siegl and E. G. Fraser, "Serum 25-hydroxyvitamin D status of vegetarians, partial vegetarians, and nonvegetarians: The Adventist Health Study-2." *The American Journal of Clinical Nutrition*, 89(5), 2009: 1686S–1692S. doi:10.3945/ajcn.2009.26736X
45 C. Koebnick, C. Strassner, I, Hoffmann, and C. Leitzmann, "Consequences of a long-term raw food diet on body weight and menstruation: Results of a questionnaire survey." *Annals of Nutrition and Metabolism*, 43, 1999: 69–79. https://doi.org/10.1159/000012770
46 Paul Kouchakoff, "The influence of food on the blood formula of man." 1st International Congress of Microbiology II. Paris: Masson & Cie, 1930: 490–493.
47 L. B. Link and J. D. Potter, "Raw versus cooked vegetables and cancer risk." *Cancer Epidemiol. Biomarkers Prev.*, 13, 2004: 1422–1435.
48 "Top 5 worst celeb diets to avoid in 2018." The British Dietetic Association (BDA), 2017. www.bda.uk.com/news/view?id=195
49 W. Schlörmann, M. Birringer, V. Bohm, K. Lober, G. Jahreis, S. Lorkowski, A. K. Muller, F. Schone and M. Glei, "Influence of roasting conditions on health-related compounds in different nuts." *Food Chem.*, 180, 2015: 77–85.
50 Cuenca-Sánchez, 2015; Ioannis Delimaris, "Adverse Effects Associated with Protein Intake above the Recommended Dietary Allowance for Adults." *ISRN Nutrition*, 2013(126929), 2013. doi:10.5402/2013/126929
51 Dhandevi Pem and Jeewon Rajesh, "Fruit and vegetable intake: Benefits and progress of nutrition education interventions-narrative review article." *Iranian Journal of Public Health*, 44(10), 2015: 1309–1321.
52 Robert Ladouceur, Paige Shaffer, Alex Blaszczynski and Howard J. Shaffer, "Responsible gambling: a synthesis of the empirical evidence." *Addiction Research & Theory*, 25(3), 2017: 225–235.
53 Alexandra Deabler, "Arby's creates 'megetables' in response to fake meat trend: 'Why not meat-based plants?'" *Fox News*, June 26, 2019. www.foxnews.com/food-drink/arbys-megetables-response-fake-meat-trend

6 Education and abolition

In Chapter 2, I argued that veganism is based on the notion that consumption of animal products is less than virtuous: that is, it is self-indulging, animal agriculture undermines human health, degrades the environment, and promotes un-aesthetic values and violence. The negative global impact of animal-based diets upon the environment and human health are alone serious enough issues. Many people are still uninformed, but many are informed about environmental issues, but choose to ignore and perpetuate their diets at the cost of environmental degradation, animal and human suffering. Worse, uninformed physicians recommend their patients animal products for good health. Although from time to time reports emerge that veganism is on the rise, arguably at present veganism has become a trend that runs the risk of riding along other trends. The uncomfortable truth is that animal-based food production is increasing.[1] The world in which we live is facing tremendous environmental issues due to the fact that meat eaters enjoy animal-based diets regardless of the global effects that these diets have upon the environment. At the same time, the clock is ticking and the environment is progressively worsening. All the good philosophical arguments offered to show humanity the gravity of the situation, and convince the public to transition to vegan diets, have not been very successful, at least not as successful as one might home. I have the feeling that there is little hope that they ever will. Thus, it is now time to make truly radical changes and consider ways to legally ban the production and the consumption of animal products.

Human food should be conducive to good health and should respect the environment. It is interesting to see that countries with a larger gap between rich and poor cause more harm to the environment. In a 2016 report, Oxfam found that "The average footprint of the richest 1% of people globally could be 175 times that of the poorest 10%."[2] In more economically even affluent countries, such as South Korea, Japan, France, Italy and Germany, the average pollution is much lower. In other words, people in more economically balanced countries consume less, produce less waste, and cause less pollution. One of the problems is that there is a positive relationship between wealth and the consumption of animal-based diets. Due to the decline in prices, developing countries buy more animal-based products.[3] However, the trouble is that those people who consume more meat affect their health and the health of those who don't eat meat, and cause a tremendous negative impact on the environment. It is plausible to say then that consumption of animal products must not be promoted—indeed livestock production must be either reduced drastically or, better yet, discontinued. Two possible routs to consider are an increase in price and a legal ban on animal-based food where consumption of this is not sustainable and equally nutritious food is available.

Price increase

Many writers, most notably Jan Deckers, have suggested that in order to reduce the consumption of animal products it is necessary to increase the price of such products.[4] Also, Robert Goodland argues that it is necessary that governments discontinue subsidies and introduce heavy taxation for the least sustainable forms of agriculture.[5] In other words, the more environmental damage a form of agriculture causes, the heavier the tax it is applied to it. Government initiatives must be implemented to reduce ecological footprint. At the same time, governments should give subsidies to encourage activities that reduce ecological footprint. Deckers refers to this possible legislative move as the "vegan project". However, it is not a *vegan* project. Rather it should be referred to as the *human* project. We must consider that the premise here is that the environment is suffering as a result of humans' irresponsible food choice. Thus, penalizing the livestock sector is not a vegan project; rather, it stems from a claim of the right to healthcare, and moreover the right to have a healthy environment. Looking at studies, it is not hard to imagine which activities are conducive to a healthier environment—planting trees, cleaning waters, growing crops organically, promoting polyculture; and which exacerbate environmental degradation—animal agriculture.[6]

One point about increasing prices and applying heavier taxes to animal-based products is that such a measure should be considered as a two-step action. What I mean is that it is plausible to think that no matter what, meat eaters will be willing to pay higher and higher prices for animal products. Also, it is undesirable and inconceivable to expect to switch to veganism straight away. A plausible way to go about it is to slowly phase out the production of animal products and eventually factory farming. The idea is to gradually increase the prices of animal products leading toward a total ban. Meanwhile, governments and industries will have the time to rethink and organize sustainable agricultures, creating jobs for those who worked in the livestock sector.

Legal ban on the production and sales of animal products

In affluent societies, consumption of animal products endangers the environment and human health unjustifiably—even the health of those who do not consume animal products. One critical issue is that of zoonotic diseases, i.e., infectious diseases that are spread between animals and people, of which the livestock sector is a leading cause.[7] Furthermore, animal-based diets are not efficient. Raising animals for food involves deforestation to grow crops and create pasture. Farm animals consume most of the grains, corn, legumes cultivated. These are foods that could directly feed humans. In other words, natural resources could be used more efficiently to produce food for human consumption.

As the world population is growing, demand for animal products is also growing. To satisfy this demand, more animals must be brought into existence. In 2018 in the US alone, over fifty billion animals were killed for food. This number is already frightening, and it reflects only the US. Obviously, more animals alive means more natural resources, such as water, fossil fuels, food, electricity, and more that have to be used. Confinement of farm animals inevitably leads to infectious diseases that are spread farther and farther since animals are transported around the world. To fight diseases, the livestock sector uses drugs, such as antibiotics, to prevent diseases. Globally, half of the antibiotics that are produced are used to prevent diseases. This promotes drug-resistant strains of bacteria.[8] Not to mention that these drugs are consumed by the animals and end up not only in the bodies of those who will eat the animals compromising their health, but also in the soil and the waters and polluting them.

Speaking of diseases, vector-borne diseases are caused by infections transmitted to people by insects. Such diseases are caused by and became more severe as a result of the environmental changes that resulted from the practices of animal agriculture, such as deforestation and reduction of biodiversity. For example, forest clearance in the early 1960s in Bolivia to create agricultural land caused the migration of mice that concentrated and overpopulated other areas carrying along a viral fever known as Machupo, which killed one seventh of the population.[9] A similar epidemic also happened in Argentina. Most diseases, HIV, influenza, the Nipah virus, are vector-borne diseases. The bottom line is that those who consume animal products contribute more to the emergence of zoonotic and vector-borne diseases that cause illness and kill people and animals than those who consume plant-based diets.

As I mentioned, another problem is that the animal agriculture sector uses too much agricultural land. What I mean by too much is that 70% of earth's arable land is used to grow food to feed animals, while many people in the world are malnourished or starving.[10] Most people are afraid that if the world were to discontinue the production of animal products there would not be enough food to eat. But in fact, only one third of arable land is used to grow food for humans. What does all this mean? It means that using land to feed animals is highly inefficient. Vegetarian diets generally require five times less arable land than meat-based diets. Consequently, meat-based diets contribute more to land degradation than plant-based diets. Thus, releasing arable land from the livestock sector will enable polyculture and the production of greater quantities of plant-based food. At present, the soil is depleted of minerals as a result of intensive animal agriculture; farmers apply phosphorus fertilizers to supplement the low quantities available in the soil. In many cases this has led to the buildup of phosphorus in the soil, and consequently the potential for phosphorus to become soluble. Dissolved phosphorus is transported from farms to lakes, rivers, and streams causing excessive aquatic plant growth, such as eutrophication (rapid growth of algae). Decomposition of algae leads to hypoxia in rivers and seas, which causes suffocation of aquatic ecosystems. Eutrophication also generates Pfiesteria Piscicida, literally a group of fish-killer eukaryotes.

Furthermore, the livestock sector uses more fresh water than any other sector. Animals consume more water than humans. This sector also pollutes more water than any other sectors. Furthermore, farmers use fertilizers and pesticides that cause the formation of nitrates that leak into the groundwater resulting in negative health effects. Farmers in the USA use recombinant bovine somatotropin (rBTS), hoemones that pollute waters. Its use is prohibited in Europe and many other countries. As a result of these pollutants, aquacultures are negatively affected. Consider that half of the fish that humans consume are produced in those aquacultures systems.

Animal agriculture's use of land is not efficient because feeding animals requires a considerably large amount of arable land and plant protein. When we do the math, it is evident that a) we could feed more people and more adequately by using the same amount of plant protein that is now required to feed the animals; b) with animal farming out of the picture, we could use less arable land in a more sustainable way; and c) animal agriculture degrades more land and has a negative impact upon the environment than any other agricultural sector. Vegan diets, on the other hand, are shown to be more efficient than any other diets. They consume less water and reduce water pollution.

Animal-based diets require more fossil fuels than vegan diets. This makes the livestock sector a leading cause of climate change as a result of greenhouse gases emissions. Although studies vary on exactly the amount of responsibility, (some studies suggest that animal agriculture causes 51% of all emission of CO_2[11]) it is clear that animal agriculture contributes to the depletion of the ozone layer as a result of animal digestion's emission of methane, and a loss of earth's photosynthetic capacity through deforestation, which reduces earth's capacity to absorb carbon from the atmosphere. Thus, there is no question that eating animal has a tremendously negative environmental impact compared to vegan diets—and therefore it should be discontinued. In other words, considering the ecological footprint that animal agriculture leaves, governments and individuals who care about the right of all humans to health care have a moral obligation to make a move toward diets that do not have deleterious impacts upon human health and the environment. Since vegan diets can reduce or in some cases eliminate these environmental impacts, people and governments should encourage vegan diets and inform people of the negative effects of meat-based diets.

The practices of the livestock sector can be described as cutthroat, profit-driven, callous, absurd, and revolting. There is nothing remotely fair, just, compassionate—nothing noble, nothing that evinces good intention or good human character—with such practices. Consider that in affluent societies we have the fortune to have an abundance of all kinds of fresh fruit, vegetables, grains, legumes, and more—you may live in New York City and eat tropical fruit in December. Yet, people demand animal flesh irrespective of the negative externalities that result from animal agriculture.

Animals like chickens, pigs, and cows are confined in overcrowded and filthy quarters. Animals are overfed and develop metabolic diseases. Also, they develop infections due to the unsanitary conditions in which they live. Every decision is made in consideration of profit rather than the welfare of the animals. One example worth mentioning is the chicken farm's treatment of male chicks. Since males do not lay eggs, chicken farms kill them immediately after hatching. They are killed by means of suffocation in large plastic bags, gassed to death, or ground alive. Chickens are then hung from their feet and their heads dunked in electrified water to cause cardiac arrest. Then, they are decapitated and dunk in boiling water. Those that are still alive drown in the water. Virtually all people (except the deranged) would be appalled at such practices. I will spare the reader from the details of slaughtering cows and pigs or other so-called farm animals. Somehow the fact that they are larger animals, with more flesh, entrails, and a lot more blood than chickens, makes their slaughtering even more gruesome. Some people, however, argue that it is possible to kill and eat animals when it is done in a humane, compassionate way. Such arguments never cease to baffle me. I wonder how it is actually possible to be humane and compassionate and at the same time decapitate, gut, bleed out, and cut up any animal (human or non-human). To be humane and compassionate means to act with benevolence and consideration. When we deal with our families and friends we act humanely and compassionately. That is, we take positive action to make sure that our loved ones are well and have a pleasant life. It would be absurd to suggest that one could be humane and show compassion toward a friend and in the end kill the friend to sell his flesh for food or his skin to make lampshades. And there is no reason as far as one can see that when it comes to animals somehow it is actually possible to treat them humanely, to show compassion, and then kill them and eat their flesh. This would not be humane treatment, but rather it would be an unfair and partial treatment.

Despite these considerations, some meat eaters and vegetarians may remain unconvinced about the moral necessity to adopt veganism. However, having documented the negative impacts of animal-based diets upon the environment and human health, and the cruelty and production of un-aesthetic values caused by the livestock sector, it is clear that diets should not be a matter of taste or personal preference. Something must be done to try to revert the damage done to the environment. I offer a valiant answer to this problem, and that is, political and legislative reforms to ensure that people will fulfill their duties to protect the environment when they make choices about what to eat. In other words, the next step is to ban the consumption of animal products, which is what I call the humane project. This is of course a colossal difficulty in light of many factors, the most serious of which is that society is animal-product-centered. We have been disciplined by the livestock sector into believing that consuming animal products is necessary, that it is normal, and that being vegetarians or vegans is a radical and unnecessary position. It is not difficult to understand why this is the case—convenience and profit.

A legal ban is an enormously complicated goal to accomplish, but not impossible, and that is the point. A ban of animal products, however, has to be the result of education. I often hear that one should decide his or her diet as an adult. This is one of the problems because it is much more complicated to convince a person who has settled way of life that his or her diet is a mistake. Therefore, I suggest certain educational reforms. For example, children since a young age should be informed of environmental issues through clear information through lectures, videos, and more, that clearly explain the impacts of animal agriculture; moral education emphasizing virtuous action; and vegan food preparation and nutrition. Instead of serving cheese sticks and chocolate milk, public schools should serve abundant fruit and plant-based products.

Talking about education immediately raises a red flag. I am not suggesting that children be brainwashed. The sort of education I have in mind is the integration of environmental studies where facts are presented in a neutral, and unbiased way.

For example, consider the following:

- Eating animals is not necessary to survive. In fact, nutrition sciences show that eating animals is deleterious to human health.
- Farm animals have horrendous lives.
- Animal agriculture is a leading cause of environmental degradation and climate change.
- Meat essentially is the decomposing, mutilated parts of animal cadavers.
- Slaughterhouses are places where thousands of animals are killed, gutted, and bled.
- Many slaughterhouse workers suffer from psychological disorder and pathological sadism.[12]

Having considered these facts, it is hard to think about any positive aspects of animal farming. Yet, people eat animals. What is the explanation? On the one hand, virtually all people regard such practices as barbaric and heartless. On the other hand, people eat animals. The explanation is very complex. Part of it is that animal agriculture, hunting, scientific research, and the entertainment business have established very powerful mechanisms to subvert and override our moral feelings of compassion and empathy toward animals. Early on in our lives, we are disciplined to regard animals as property and food. Children are fed animals in forms that do not resemble animals, things like mush, nuggets and things labeled "happy meals;" yet, they are read "Hey, diddle, diddle ... the cow jumped over the moon. And the dish ran away with the spoon," though the cow in the story is not *the* dish. In most cases, caregivers avoid discussing with children that the cow does not jump over the moon but she is slaughtered instead.

Advertisements, meat trade, and hunting journals are deliberately deceptive about animals. They all work hard to make us believe that animals exist for our benefit, that eating meat is normal, and that we must eat it. A typical expression about meat is that it is "juicy." What juice? The juice is blood! Butcher shops prefer to be called "meat markets." Slaughterhouses are "meat plants" or "meat factories." Terms such as beef, pork, mountain oysters, drumsticks, and other euphemisms are used to refer to animal flesh. Vivisectionists prefer the term "dispatch," or "sacrifice" instead of "kill." Hunting is regarded as a "sport" of "harvesting" animals.

Deliberately deceitful language about meat pervades society. TV shows, movies, and popular culture are anti-vegan. In the sit-com *Two and a Half Men*, Charlie is the stereotypical man who likes steak. In one episode, he dates a vegan ballerina, who takes Charlie to a vegan restaurant. Charlie protests that he cannot stand having to eat "medallions of bean curd in lawnmower sauce."[13] Veganism is constantly ridiculed. Heroes are never vegans or vegetarians. Heroes are men, and men are strong, and strong men eat meat, while girly girls eat salad. Vegans are portrayed as obnoxious salad eaters who bother other people, feel superior, and pretend to save the world. The very term, "vegan" is nowadays ubiquitous and is associated with a cult-like attitude. Animal exploitation thrives not because it is normal, but because people are deliberately manipulated by a system of exploitation, which comprises the media, scientific research, meat and dairy industries, hunting, and the food industry. If we are constantly told these messages, we cannot be properly informed about the lives of animals, and therefore cannot sympathize with them.

What forms of instructions, then, can inform in a non-manipulative way and help people overcome the false idea that animals are humane property and food? It is a rather complicated issue that involves multi-million-dollar industries that hold sway over government and the media. The first step is to recognize the subversive ways that society uses to discipline people out of their natural love for animals and their sense of what is just and compassionate. One answer is to become more involved in moral education and move in the direction of a virtue-oriented ethic advocating the importance of relationships, care and compassion. Viable ways of instruction will require a demand for clear and transparent information from the government and the media. This may start by educating children about the lives of animals and animal exploitation.

There are some YouTube videos of a food demonstrator in a supermarket offering samples of sausage. When customers want to purchase the product, the demonstrator removes a piglet from a box and holds it over a meat grinder. The reaction of all customers is the same: They are shocked and appalled at the cruelty and cold heartedness of the demonstrator and his lack of scruple about killing the cute piglet. This predicament occurs while most of the customers are still chewing their sausage. But then, why are the customers shocked? Why are meat eaters opposed to animal exploitation? Usually people say that they are against inhumane treatment of animals. If one has such worries, why eat animals at all? Such conflicting attitude evinces an inconsistent morality and a truncated expression of our moral feelings of compassion, fairness, and benevolence toward non-human animals. The point of moral education should be to nurture these feelings and enable us to fully express them in harmony with reason. Education in my view should embody virtues such as compassion, justice, temperance, and magnanimity. The practices of the livestock sector are in no sense consistent with these virtues. Consistency demands that we direct these feelings toward all animals, and not just those closer to us. A fair individual is one who is fair in all circumstances. The compassionate, fair-minded, and magnanimous person feels sympathy for all animals and would feel ashamed of allowing the blood, death, and squalor that farm animals experience on farms and in slaughterhouses. In other words, a virtuous individual does not regard animals as food or property, and consequently chooses to be an ethical vegan.

Further objections

Objection 1: Aren't people's food choices a matter of personal freedom?

Answer: Most thinkers, and not only modern thinkers if you consider the like of Pythagoras, realize that what we eat is a serious moral concern. Unfortunately, the activities of the livestock sector cause major harms to animals, to people's health, and to the environment. Consequently, a ban is not only desirable but also necessary to save the world from degradation.

Objection 2: An abolition of meat seems undemocratic

Answer: As many other social and moral issues, the abolition of animal-based products should be considered and discussed. But again, considering the harm that the meat industry causes to the world, it is in fact the meat industry undemocratic for perpetuating the exploitation of animals.

Objection 3: If we abolish the production of animal-based food, won't we harm an economy that produces jobs?

Answer: First, there is no concrete plan for a ban. The details would be discussed among experts in the law, policymaking, economy, and ethics. In any case, a ban would be progressive and jobs would be progressively replaced by other sectors. Second, given the gravity of the situation the livestock sector should be stopped irrespectively of other considerations. Even some scholars, on the ground that it might have a negative effect on the economy, opposed the abolition of slavery. The point still remains that slavery was immoral and had to be abolished regardless of financial considerations. It does not follow, however, that those in favor of the abolition of meat take financial considerations with a grain of salt.

Objection 4: The production and consumption of meat is essential to the sustenance of many humans

Answer: This is the very problem: production and consumption of meat is not essential at all. We have been disciplined into believing that it is so. Actually, by discontinuing the meat industry, it will be possible to release more arable land that would yield more food for humans. Once again, we have to remember that the richest courtiers in the world are the ones that consume lots of meat.

Moral education

Some people think that meat eaters should visit slaughterhouses to experience emotionally the horrific experiences that animals undergo. This is certainly a valuable recommendation. But many people do not respond well to violent images, not only because of the graphic nature of those images but also because many become indifferent and unmoved by such an experience as visiting a slaughterhouse. Nowadays, people have grown so accustomed to violence and gore, not only real but also fictional. Our tolerance of violence has been stretched by such events as people jumping out windows of the collapsing Twin Towers, the Boston Marathon bombing, the execution of innocent people in a movie theater, and more. I think the images of animals being slaughtered might work for some but not all. I've heard too many people say, after visiting a slaughterhouse, "Hey, that's life." Besides, many people are just reluctant to visit a slaughterhouse. Moreover, slaughterhouses are tucked in remote areas—precisely to avoid unwanted eyes seeing the blood and guts and their noses smelling the stench of death. Experiencing a slaughterhouse could sensitize certain individual, but not all; and the negative experience might push people further into denial. In my experience, many meat eaters have a change of heart when they have a positive experience; when they come face to face with animals, get to know them, play with them, talk to them, and realize that they are creatures that have feelings and needs like ours or very similar to ours. Then, it is possible to realize that animals are not human food. But first, I think, it is necessary to deprogram people from the false idea of food that was inculcated in them by society.

Josephine Donovan proposes a conception of care ethics.[14] Such an approach suggests that humans pay attention to the needs of animals by connecting with them, by listening to them and learning about their opinions. Minorities and marginalized people, for example, are aware of what it feels like to have their opinion ignored. This experience should make them realize that humans have marginalized animals and treated them as property and as food. Lori Gruen, who proposes assessing the issue of animal agriculture from the point of view of a woman writes, "By consuming animal bodies, women are implicitly supporting their domination."[15] There is a strong intellectual connection between veganism and feminism in that, throughout history, men have abused women just as the meat industry, along with hunting and the masculine image of eating meat, remain a male-dominated affair. Consider, for example, that female cows, turkeys, horses, who are abused emotionally and physically. Humans forcibly impregnate them.[16] Artificial insemination is the ultimate violation of an individual. Farm workers collect the semen of animals and inject it into females. This is not an example of virtue. Unfortunately, most people are not informed of such practices or, if they are, they have been disciplined to accept them and think that such practices as artificial insemination, and even slaughtering, can be done humanely.

Can the experience of caring for other individuals be universalizable? One of the greatest moral philosophers in history, Immanuel Kant argues that feelings are not reliable and the capacity for sympathy is not evenly distributed in the population. Kant made a valid point. However, at the same time, one should be wary of such criticisms because in societies those who hold political power, traditionally men, are the ones who have traditionally done the universalizing and have, also traditionally, left out women, the environment, and minorities. When we generalize from the perspective of a minority, as Donovan points out, "it is not illogical to contend that one might easily generalize from an individual ethical reaction, extending that reaction to others similarly situated, thus positing a general or universal precept."[17] When the generalizing is done by people who really know what it is like to be ignored, and is done from a care perspective, then universalization of care is possible. Karen Warren, an ecofeminist, points out, "There are important connections—historical, experiential, symbolic, theoretical—between the domination of women and the domination of nature."[18] Other writers agree with this point and come to the same conclusion.[19]

Most animals, at least those that people eat, are living creatures with whom one can communicate cognitively and emotionally and find out about their needs and wishes. It is not hard to understand animals' body language when they suffer, and generate a moral duty to care for that animal. The constant suggestion is that it would seem permissible to raise animals and slaughter them humanely as long as animals are treated with care. In fact, many vegan sympathizers and vegetarians often make this point, that is, yes, let us be against factory farming and its cruel practices, but what about the happy cow that lives in the barn and happily shares her milk with us and, after a beautiful and peaceful life, is "humanely" killed and turned into food? In my view and from an ethic of care's perspective, there is no such a thing as a "happy cow" in a barn or such a thing as "humane" killing. A happy cow is one who is left alone, and killing a being to use its flesh as food is not a sign of caring. People have domesticated cows and raised them for food for centuries. Notice also that when people make assertions such as "happy cows," they—the people and not the cows—decide whether a cow is happy or not. These oxymoronic assertions always make me wonder whether people are interested in truth or just wish to play academic games. They think that the mere addition of the word *humanely* in front of the word *slaughter* makes perfect sense; but it is no more sensible than saying darkness visible. Slaughtering is not something that could ever be humane. Most importantly, in raising an animal with care, if one does this with the intent to fatten him up and eat him, again, it is an error to say that the animal was raised with care. Caring entails love for others, but raising animals to be slaughtered and eaten is certainly not a loving behavior.

This point is frequently brought up by meat eaters who oppose the gratuitous cruelty inflicted upon animals by the meat and dairy industry, but argue that it is not an immoral practice to raise animals "humanely" for the purpose of, for example, obtaining milk or wool or even food, so long as said animals are treated "humanely." To raise an animal with care means to love and care for that animal. That is what people do with companion animals: they raise them with care, feeding them, loving them, and giving them a shelter and place in their homes. Caring means understanding creatures emotionally and valuing what is important in their lives. For animals like cows and pigs, for example, it is of great importance that they may live freely and pain free and enjoy their families and friends. Caring for an animal builds a connection based on love and friendship that makes us see an animal as a friend and not the source of food or clothing. One important aspect of becoming a vegan is that one realizes that the question of our treatment of the other animals becomes moot. According to the human diet, the question of whether consuming the milk of a "happy" cow is moral or immoral becomes irrelevant. One realizes that a cow is a peaceful animal whose milk is in no way human food, but rather food for her babies. By the same token, even if a cow has milk "left over" after she has fed her babies, her milk is still not human food.

The importance of dialogue and moral education cannot be overemphasized. To make significant progress in morality, in general, and in our understanding of our relationship with animals, in particular, we ought to construct a dialogical ethics to determine the treatment of other beings based upon their needs. If they are human beings, we can clearly understand what they want. If they are animals, we can relate with them and understand them. If we don't, we must learn the language of nature. Early on in our lives, our parents and society condition us to detach ourselves emotionally from animals and deny the reality of what goes onto our plates. Children have to be educated out of the early sympathy they feel for animals. Indeed, in order to eat the bodies of animals, wear their skin, hunt them for fun, and allow useless torture disguised as "scientific research," people are disciplined to do so by manipulative means, such as deceitful language that praises animal exploitation and often ridicules ethical veganism. Bodybuilders are told to eat meat to grow muscles, while eating a vegetarian diet is typically denigrated, "Tofu just might kill u tossed salad makes u weak," writes Ted Nugent on his Twitter page.[20] Parents feed children animal products from a very young age, without giving them a choice in the matter and without educating their children about what they are eating; parents tell them about "happy cows," and "happy meals" that deliberately conceal the fact that those meals come from the exploitation and suffering of animals—the same animals, by the way, children love to see or read about in fables. As Carol J. Adams notes,

> We live in a culture that has institutionalized the oppression of animals on at least two levels: in formal structures such as slaughterhouses, meat markets, zoos, laboratories, and circuses, and through our language. That we refer to meat eating rather than to corpse eating is a central example of how our language transmits the dominant culture's approval of this activity.[21]

The first objection is that people do not want to give up consuming animal products.[22] They are not ready to go vegan, and consequently it would more fruitful to think a way to continue producing animal products but in ways that take into account environmental interests. In fact, it is quite the opposite. There is evidence that people are ready to make changes. Anecdotally, as I mentioned earlier, it may be observed the recent interest in veganism. Non-philosophers have become more and more interested in veganism because they understand that animal agriculture contribute to the degradation of the environment; that eating more fruit and vegetables is more conducive to good health, which is a no-brainer that somehow has been contested, not surprisingly, by the meat industry; and that meat-based diets require the unjustified infliction of pain to farm animals. I would only add that humans are, in fact, not clamoring meat and animal products. Most people eat what they can and what they are told to eat. Furthermore, it is evident that veganism as an ethical and dietary movement is growing. As a recent article in the *The Economist* suggestively states, the year 2019 is "The Year of the Vegan: Where millennials lead, businesses and governments will follow."[23]

Most people (except for some hunting enthusiast or some peculiar individuals) detest the sight or even the idea of blood, dead animals, and everything that is associated with slaughtering. There is anecdotal evidence, of course, of people who won't eat their steak unless is well-done because they can't stand the blood or people who eat fish but refuse to look at the whole fish prior to cooking it. Also, many meat eaters like to eat meat but are squeamish about handling raw meat. Meat is cold, dead flesh. Many of us find the feel of raw meat just too close to the real thing for comfort. There is also scientific evidence. For example, consider the question of taste. In 2016, two researchers, Anderson and Feldman Barrett, tested how people's beliefs of how animals are raised can influence their experience of eating meat. People were given samples of meat, each of which was given full descriptions of its origins and the treatment that the specific animal received on the farm. Some samples were said to be from factory-farmed animals, while others were labeled as "humane". In reality, however, all the samples were identical. Fascinatingly, the participants of this study experienced the taste of the samples differently: meat described "factory farmed" was perceived as looking, smelling, and tasting as less pleasant than "humane" meat. The difference was even to the degree that factory farmed meat was said to taste more salty and greasy than "humane" meat. Furthermore, the participants who were told that they were eating factory-farmed meat consumed less of the sample. According to the authors of this study, "These findings demonstrate that the experience of eating is not determined solely by physical properties of stimuli—beliefs also shape experience."[24]

People consume meat because the meat has been disassociated with those negative aesthetic values, violence, suffering, blood, confinement, and so on. Why, then, do people consume meat? There are many reasons to me mentioned. An obvious reason is that animal products are everywhere. Especially affluent societies in the world overemphasize meat and animal products. This is because corporations are trying to shove animal products down the throats of the public, for reasons that have nothing to do with health or morality. I believe that production is prior to consumption. It is not the consumers who create demand for products; rather, it is those who own the production system that make decisions for the public—the music they should listen to, how to dress, what to buy, and the food they should eat. The point here is that there is no inherent or natural reason that people should want to eat animal products. Eating animal products is like smoking cigarettes or using gasoline as a combustible—they produce them, we consume. In other words, animal products are just part of a long list of things that people consume because the market gives no serious options but to consume those products. The meat industry, magazines, doctors, TV shows, the health industry all sing in unison the song of meat—meat is good for you, buy it and eat it. Consequently, information and education alongside more options in the way of plant-based foods will likely facilitate the legal ban of animal products.

Most advertising contains a very persuasive component that manipulates people into buying products. Unfortunately, advertising is not the business of making consumers have an understanding of the features of the products for sale. And since advertising is not truthful, people are not free to choose the right products. The issue here is not whether deceptive advertising is wrong—most believe it is. As I discussed earlier, the common notions of food, diet, health, and masculinity are distorted by the livestock sector and by the powers that have an interest of selling animal products at stake. The meat industry and its advertising does not inform people how to acquire what they want, but rather gives people new wants. Moreover, since we are bombarded with advertising and messages about consuming animal products, people want too many of those goods and not enough public goods or moral goods.

The animal-products industry implement an aggressive form of associative advertising, by associating animal products with a positive belief, feeling, attitude, or activity that usually has little to do with the product itself: Need omega-3s? Eat fish. Need calcium? Drink milk. Need proteins? Eat beef. The reality is that a diet of fruit and leafy greens can provide all those nutrients and in a more efficient and ethical way. The sort of advertising and information regarding animal products is thus misleading and sanitized from moral values and environmental considerations. This sort of advertising attempts to create desires in people by subverting their autonomy.

What I suggest is the human diet—a diet of fresh, unprocessed, uncooked, fruit and leafy greens, with the occasional nuts and seeds. The human diet, unlike other forms of veganism, can reduce environmental degradation and the systemic domination and mass violence to an anthropocentric and self-indulging personal ethics, consumption, and intemperate attitude toward food. Our focus, I argue, needs to be on us insofar as a quest for living a good life through the acquisition of fundamental virtues; moreover, we should redirect our focus on the victims of our violence, the animals, and our environment. Veganism is understood by most of the public as a position about diet, not a moral position that opposes violence and the destruction of the environment. Veganism, unfortunately, in the reality of the everyday world is compatible with speciesism, scientific exploitation of animals, circuses, and habitat destruction. The human diet, on the other hand, focuses on them and thus is the most ethically sound approach to nutrition. At any rate, in light of the foregoing analysis, it is absurd to continue to structure our civilization around flesh. Most people—including meat eaters—recognize this absurdity, but have their hands tight up. Consequently, a legal ban on animal products can be accomplished.

The second objection is that a legal ban would seem to undermine human food security. This worry seems groundless especially considering that discontinuing the livestock sector would release more arable land and it would allow more biodiversity and a greater abundance of plant-based food. The third objection is that a legal ban on animal products may alienate human beings from nature. In my view, this is quite an extravagant worry. First, there are many human endeavors that have alienated us from nature. I am reminded of that every time I go to work on an overcrowded train where every single person stares at his or her cellphone, holding a cup of coffee in the other hand. It is hard to see how the perpetuation of factory farming and killing animals will bring us closer to nature. What I propose here is the human diet, a diet that is specific and optimal to humans. Consequently, the prospect of eating fruit and vegetables instead of dead animals, far from alienating humans, will bring humans back to nature and to their nature.

Another objection is that veganism promotes consumerism. This is not entirely false. That is one of the main reasons why I am promoting a very specific type of ethical veganism, which is the human diet. Veganism, one may note, is very fashionable these days. As a result, many non-vegan companies, such as Burger King, McDonald's, White Castle, Tyson, and more, are now jumping into the business of plant-based meat. Consequently, by purchasing and consuming these products, sadly, and ironically, vegans would support non-vegan companies that promote the livestock business. In a way, many companies that support the meat industry are appropriating veganism, which is reducing veganism to an act of consumption and is obscuring the real moral motivations behind veganism. Thus, I propose the human diet as the best approach to veganism. To adopt the human diet is to reconfigure our relationship to animals and nature from exploitation and supremacy to love, compassion, and solidarity. The human diet, moreover, promotes a more humane and creative existence by returning to a simpler way to approach food and abandoning our social dependency on the products of animal exploitation. The human diet enables us to develop and rediscover our natural food; to use less natural resources, such as gas, electricity, water; to buy and use detergents, disinfectants, and all the (some might say) superfluous and expensive appliances that are used to prepare cooked food. In other words, the human diet, a specific kind of ethical veganism, is a form of resistance to consumerism.

Yet another objection is that any talk of human diet has to come to terms with the fact that many people have no access to healthful food. The problem with this sort of criticism is that it does not consider the fact that all systems of exploitation and injustice are interwoven. They arise from the same mentality of dominance and power that the philosophy of the human diet opposes. We cannot even begin to hope for justice and allocation of healthful food for all humans if we continue to perpetuate a system of violence toward animal and nature. The idea of the human diet is to overcome unequal resource distribution. The goal of the human diet is to advance food justice by developing and redirecting subsidies from ecologically destructive agriculture toward a sound one that frees up land to grow abundant fruits and vegetables. This is important as an example to other nations to inspire campaigns against the livestock industry, which is interested in profit at the expenses of global health. By moving in this direction, we can hope to revert the damage done to the environment.

Some also point out that veganism is a luxury. Again, as I discussed above, there is a sense in which veganism has become a business rather than an ethical position and endeavor. And, once again, this is another important reason for promoting the human diet. We have to consider, however, that eating animal products is an antonomasia of first world luxury. Animal products rely on a globalized food system of colonization of land, deforestation, domination, and pollution. The livestock sector's workers are employed in jobs of mass killing of billions of suffering animals. We have to keep in mind that without the exploitative nature of industrial agriculture and global capitalism, it would not be possible to have the amount of animal products consumed by millions of people who live in affluent societies. It is, after all, very telling that the world's richest countries eat the most animal-based products, while in the poorest countries around the world people survive eating plant-based diets. The human diet is the most practical way to stop exploiting the land and poor people to feed the wealthy animal products. A raw vegan diet uses natural resources and the land for the benefit of the environment, the animals, and impoverished communities around the world.

Veganarchism

Anarchism, in a nutshell, argues that the state and its laws are unjust because they are founded on coercion.[25] Capitalism, exploitation, and alienation are, so to speak, built-in in the very idea of a government. Thus, most social issues, such as discrimination, poverty, and more, are the inherent in the state. Consequently, anarchism is for the abolishment of the state. A typical misconception is that anarchism is synonymous with immorality, chaos, bob-throwing and the like. On the contrary, most anarchists in proposing a stateless society emphasize cooperation among people and acknowledge the importance of virtues such as justice, compassion, friendliness, and care. The idea of anarchism as an ideology consistent with virtue ethics, or at least consisting in principles compatible with virtue, exists though it is not widely explored. Benjamin Franks, for example, argues that anarchism embodies many traditional virtues.[26] Interestingly, many anarchists, who believe that anarchist ethics is consistent with virtue ethics, come to the (to me obvious) conclusion that animal exploitation is immoral. Another example is Élisée Reclus, who in the early twentieth century wrote a defense of vegetarianism. Once again, we see the same rational conclusion that our treatment of animals is connected to our desire to dominate and exploit.[27]

Brian Dominick, a veganarchist, also proposes that there is an interesting connection between anarchism and veganism, a movement he calls "Veganarchism". He explains veganarchism as follows:

> I am vegan because I have compassion for animals; I see them as beings possessed of value not unlike humans. I am an anarchist because I have that same compassion for humans, and because I refuse to settle for compromised perspectives, half-assed strategies and sold-out objectives. As a radical, my approach to animal and human liberation is without compromise: total freedom for all, or else.... Any approach to social change must be comprised of an understanding not only of social relationships, but also of the relationships between humans and nature, including non-human animals.[28]

Dominick, among other veganarchists, agrees with those care ethicists on the notion that there is a connection between the exploitation of animals and the oppression of marginalized groups of humans. Dominick argues that racism, sexism, and other forms of discrimination stem from the same roots as oppression of animals. He adroitly states: "To decide one oppression is valid and the other not is to consciously limit one's understanding of the world; it is to engage oneself in voluntary ignorance, more often than not for personal convenience"[29] He also worries about the appropriation of veganism by a capitalist system, and that many people nowadays worry about the suffering of animals, and rightly so, but do not see it as a problem of domination and exploitation which is consistent with capitalism. He writes, "Many vegetarians fail to see the validity of human liberation causes, or see them as subordinate in importance to those of animals who cannot stand up for themselves."[30]

The point that I argue is that to the extent that veganism is in part a change in consumption patterns, consumption alone is not likely to make the changes to end animal exploitation. Consequently, raw veganism, the human diet, is the natural and most sensible outcome of a resistance to domination and exploitation. Thus, veganarchists believe that they ought to not only fight to change our current domination of non-human animals, but should also simultaneously begin to build an alternative society characterized by different relations with non-human animals.

Notes

1 Hannah Delvin, "Rising global meat consumption 'will devastate environment'." *The Guardian*. www.theguardian.com/environment/2018/jul/19/rising-global-meat-consumption-will-devastate-environment
2 Oxfam. "Extreme Carbon Inequality." https://www-cdn.oxfam.org/s3fs-public/file_attachments/mb-extreme-carbon-inequality-021215-en.pdf
3 World Health Organization. "Availability and changes in consumption of animal products." 2015. www.who.int/nutrition/topics/3_foodconsumption/en/index4.html
4 Jan Deckers, *Animal (De)liberation: Should the Consumption of Animal Products Be Banned*. London: Ubiquity Press, 2016.
5 R. Goodland, "Environmental sustainability in agriculture: Diet matters." *Ecological Economics*, 23,1997: 189–200. doi:http://dx.doi.org/10.1016/S0921-8009(97)00579-X
6 Brad Plumer. "Study: Going vegetarian can cut your food carbon footprint in half." *Vox*, 2016. www.vox.com/2014/7/2/5865109/study-going-vegetarian-could-cut-your-food-carbon-footprint-in-half
7 Jan Deckers, 2016: 17–21.
8 T. F. Landers, B. Cohen, T. E. Wittum and E. L. Larson, "A review of antibiotic use in food animals: Perspective, policy, and potential." *Public Health Rep.*, 127(1), 2012: 4–22. doi:10.1177/003335491212700103.
9 Global Environmental Change and Human Health. *Science Plan and Implementation Strategy*. 2019.
10 H. Steinfeld, P. Gerber, T. Wassenaar, V. Castel, M. Rosales and C. de Haan, "Livestock's long shadow: Environmental issues and options." Rome: FAO, 2006. www.faostat.fao.org (accessed January 19, 2017).
11 R. Goodland and J. Anhang, "Comment to the editor." In Herrero et al. "Livestock and greenhouse gas emissions. The importance of getting the numbers right." *Animal Feed Science and Technology*, 172, 2012: 252–256. doi:http://dx.doi.org/10.1016/j.anifeedsci.2011.12.028
12 Michael Lebwohl, "Call to action: Psychological harm in slaughterhouse workers." *The Yale Global Health Review*, 2016. https://yaleglobalhealthreview.com/2016/01/25/a-call-to-action-psychological-harm-in-slaughterhouse-workers/
13 E. Gorodetsky and M. Roberts, My Tongue is Meat, episode from *Two and a Half Men*. Los Angeles: CBS, 2006.
14 Josephine Donovan, "Feminism and the treatment of animals: From care to dialogue." *Signs*, 31(2), 2006: 306
15 Lori Gruen, "Empathy and Vegetarian Commitments." In *The Feminist Tradition in Animal Ethics*, Columbia University Press, 2007: 333–344.
16 I. Gordon, *Reproductive Technologies in Farm Animals*. CABI, 2004.
17 Donovan, 2006: 308.
18 Karen Warren, "The promise and power of ecofeminism." *Environmental Ethics*, 12 (2), 1990: 126.
19 Andree Collard and Joyce Contrucci, *Rape of the Wild: Man's Violence Against Animals and the Earth*. Bloomington: Indiana University Press, 1989.

20 Ted Nugent, Twitter Post, January 12, 2014. https://twitter.com/tednugent/status/ 422456432 856809472
21 Carol J. Adams, *The Sexual Politics of Meat: A Feminist-Vegetarian Critical Theory*. Bloomsbury Academic, 2015: 47.
22 The following three objections are also addressed by Deckers, 2016: 118–128. I add further comments to them.
23 John Parker, "The year of the vegan: Where millennials lead, businesses and governments will follow." *The Economist*, 2019 (accessed September 3, 2019).
24 E. Anderson and L. Barrett "Affective beliefs influence the experience of eating meat." *PLoS ONE*, 11(8), 2016: e0160424. doi:10.1371/journal. pone.0160424 www.unep. fr/shared/publications/pdf/dtix1262xpa-priorityproductsandmaterials_report.pdf
25 Errico Malatesta, "Towards anarchism." Marxists Internet Archive. www.marxists. org/archive/malatesta/1930s/xx/toanarchy.htm
26 B. Franks, Anarchism and the virtues. In Franks B. and Wilson M. (eds) *Anarchism and Moral Philosophy*. Palgrave Macmillan, London, 2010.
27 Élisée Reclus, "On Vegetarianism." *TheAnarchistLibrary.org*, 2009. https://theana rchistlibrary.org/library/elisee-reclus-on-vegetarianism
28 Brian A. Dominick, *Animal Liberation and Social Revolution: A Vegan Perspective on Anarchism or an Anarchist Perspective on Veganism*. Syracuse: Critical Mass Media, 1995: 5.
29 Dominick, 1995: 10.
30 Dominick, 1995: 8.

7 Raw veganism and children

In this chapter, I discuss some aspects regarding veganism and health. Specifically, respond to the challenge that vegan diets can be nutritiously inadequate, and harmful to the one's social well-being, with an emphasis on vegan children. My objective in this chapter is to show that the notion that vegan diets are nutritionally deficient is a confused and groundless notion; also, I present compelling evidence that veganism in no way threatens the social lives of vegan children. Moreover, I want to break the myth that being or becoming a vegan requires sacrificing one's social life, careful planning, and supplementation. My discussion will follow an argument recently proposed by Marcus William Hunt that vegan parents have good reasons for *not* raising their children on a vegan diet because veganism is potentially harmful to children's physical and social well-being. In my rebuttal, first I show that in practice all vegan diets, with the exception of wacky diets, are beneficial to children's well-being (and to adults' well-being for that matter); and that all animal-based diets are potentially unhealthful. Second, I show that vegan children are no more socially outcast than any other group. In other words, veganism does not harm the lives of children. Having considered several studies, I show that the moral reasons that vegan parents may have for raising their children on a vegan diet significantly outweigh the reasons for raising their children on an animal-based diet. Thus, I conclude that parents have a moral obligation to raise their children on a vegan diet.

Should vegan parents raise their children on an animal-based diet?

Marcus William Hunt thinks that the answer to the question posed in the title of this section is affirmative. He argues that parents have good reasons for not raising their children on a vegan diet because "a vegan diet bears a risk of harm to both the physical and the social well-being of children."[1] Hunt argues that the moral reasons that he suggests are as strong as those that vegans have for raising their children on a vegan diet. Hunt's argument is useful because it encompasses the typical worries and reservations that people have about veganism. What I show here is that: 1) the assertion that a vegan diet bears any physical and social risk to children is a misconception. In fact, vegan parents should raise their children on a raw vegan diet or possibly a prevalently raw vegan diet; 2) the moral reasons that vegan parents may have for raising their children on a vegan diet outweigh the reason for raising their children on an animal-based diet; and 3) vegan parents have strong reasons generating a moral obligation to raise their children on a vegan diet.

Hunt's argument is the following (I paraphrase):

1   Raising children on a vegan diet can harm children physically and socially.
2   Parents have good moral reasons to avoid harm to their children's physical and social well-being.
3   Therefore, parents have good reason to not raise their children on a vegan diet.[2]

The question is, "do the offered moral reasons give vegan parents a Justification in not raising their children on a vegan diet?" To answer this question first I would like to make an important observation. Many people speak generically about "a vegan diet." Veganism is not in principle a diet, but rather an ethical view according to which eating animal-based food, using animal byproducts, and other forms of animal exploitation should be shunned as much as possible. Since the only aspect in common to vegans is their avoiding animal-based food, there isn't one specific vegan approach to diet, but many. One problem in particular is that these types of arguments against "vegan diets" purport to show that a vegan diet bears the risk of damaging vegan children's health and social lives; however, these arguments typically lack examples of vegan diets that bear such a risk. Hunt concedes that an "appropriately" planned vegan diet can meet all the nutritional requirements. In fact, the American Dietetic Association and the Academy of Nutrition and Dietetics show that vegan diets are appropriate for all stages of the life cycle, they can prevent and even reverse many diseases,[3] and are of a high nutritional quality.[4] His worry is that since not all vegans follow an appropriately planned vegan diet, they run the risk of harming their children's health. What is an example of a nutritiously deficient diet? What is an example of an "appropriately" planned vegan diet that satisfies nutritional requirements and one that does not? These types of arguments that seek to show potential risks of following vegan diets eventually collapses on themselves.

When a diet is referred to as vegan, it is not clear in what that diet consists. For example, drinking champagne for breakfast, eating French fries and vegan chocolate for lunch, and marshmallows, beer, and scotch for dinner is a vegan diet. Such a diet would likely lead to some health problems. However, what sort of parent would raise a child on a diet like that? Most importantly, assume now that animal products were added to the aforementioned diet. Would the diet be any more healthful? It seems clear that it would not. Any diet must have some kind of planning. But planning denotes something complicated, though eating a healthful vegan diet is not complicated at all, especially if it is a raw vegan diet. A poorly planned vegan diet may be just as bad as a poorly planned meat and dairy diet. In other words, animal products do not have magical power that make a diet automatically healthful. Most meat eater eat a diet that looks like this: bacon, cheese, eggs, milk for breakfast; some kind of meat and starch for lunch, and some kind of meat and starch for dinner (not to mention snacks in between meals). Diets such as this are not healthful in any respect.

According to the Office of Disease Prevention and Health Promotion's dietary guidelines 2015–2020, "The typical eating patterns currently consumed by many in the United States do not align with the Dietary Guidelines."[5] To say that eating patterns in the US "do not align with the dietary guidelines" is a very mild way to put it when we consider that, "About three-fourths of the population" consumes a low amount of fruit and vegetables. Also, "More than half of the population is meeting or exceeding total grain and total protein foods recommendations, [and] are not meeting the recommendations for the subgroups within each of these food groups." In particular, "most Americans exceed the recommendations for added sugars, and saturated fats."[6] And it must be borne in mind that saturated fats come mainly from animal sources, including meat and dairy products. Furthermore,

> the eating patterns of many are too high in calories…. The high percentage of the population that is overweight or obese suggests that many in the United States overconsume calories … more than two-thirds of all adults and nearly one-third of all children and youth in the United States are either overweight or obese.
>
> (Para. 1)

My point here is simple, in most cases eating meat-based diets is harmful. It is meat eaters than tend to not consume deficient diets. Raw vegans, by very definition, consume an abundance of fresh fruit and vegetables. Thus, the suggestion that only vegans must be careful planning their diets is inaccurate because all people, regardless of their diet, must plan well what they eat.

Hunt is worried about the appropriateness of vegan diets, "However, when we shift from talking about the in principle appropriateness of well-planned vegan diets to talking about the actual dietary practices of vegans" we see that premise 1 is true or plausibly true.[7] Hunt implies that following a well-planned vegan diet is quite difficult or that a well-planned vegan diet requires special planning. The diet I propose is not complicated at all. It requires only eating an abundance of fruit, leafy greens, and some seeds and nuts.

Health concerns

Many people are concerned about getting certain nutrients and are afraid that vegan diets cannot provide them. Hunt, for example, writes that

Vegan diets that are poorly planned are widely acknowledged to be liable to deficiencies of vitamins A, B12, D, iodine, calcium, iron, and various fatty acids such as eicosapentaenoic acid (EPA) and docosahexaenoic acid (DHA), as well as creatine and taurine (Cofnas 2018).[8]

These assertions are inaccurate. Going over all the micronutrients mentioned here is quite a task; thus, I will try to briefly describe what is inaccurate about Hunt's reading of the science. Take vitamin A for example. First of all, in determining the amount of a specific nutrient, it must be taken into account the question of absorption. For example, a US Department of Agriculture article titled "B12 Deficiency May Be More Widespread Than Thought" states the following,

> Nearly two-fifths [40%] of the U.S. population may be flirting with marginal vitamin B12 status ... the researchers found no association between plasma B12 levels and meat, poultry, and fish intake.... It's not because people aren't eating enough meat. The vitamin isn't getting absorbed.[9]

Secondly, vitamin A is not exclusive to animal products. It is found in optimal amounts in a variety of ordinary fruit and vegetables, such as potatoes, broccoli, cucumbers, apples, corn, bananas, and many more. In fact, it would be almost impossible to suffer from vitamin A deficiency on any vegan diet. The key point here is that premise 2 of Hunt's argument states that "Parents have pro tanto moral reason to not make their child bear a risk of harm to their well-being." In other words, the assumption is that parents concerned about the well-being of their children make sure not to raise their children on diets that can cause harm. Such parents, presumably, are informed enough to know the very basic rules of thumb of nutrition, which are very simple, i.e., consume fresh fruit and vegetables in abundance. There is a reason why the proverb goes, "An apple a day keeps the doctor away" and not "a steak a day ..."

No vegan parents concerned about the well-being of their children would raise their children on beer and fritters. By using the term "well-planned," Hunt suggests that vegans must plan their diets like astronauts must plan their space meals. A well-planned vegan diet does not consist of exotic or hard-to-come-by food. It is quite simple to follow a well-planned vegan diet because it is a diet rich of fruit, vegetables, grains, and legumes. Hunt seems to imply that meat eaters automatically get all essential nutrients, but vegan must be very careful. But then how do we explain that meat eaters suffer from vitamin A deficiency? It is quite clear from scientific evidence that vitamin A deficiency is in no way endemic among vegans. According to Higdon,

> In developing countries, vitamin A deficiency and associated disorders predominantly affect children and women of reproductive age. Other individuals at risk of vitamin A deficiency are those with poor absorption

of lipids due to impaired pancreatic or biliary secretion and those with inflammatory bowel diseases, such as Crohn's disease and celiac disease.[10]

In other words, the scientific literature recognizes that vitamin A deficiency are prevalent among people in developing countries, and especially people who suffer from certain diseases that prevent absorption of vitamin A. There is no evidence that vegans are more susceptible to vitamin A deficiency than meat eaters, unless they are malnourished; but in such a case malnourished is the culprit and not a vegan diet.

Regarding B12 deficiency, as discussed above the first point is absorption. Vitamin B12, also known as cobalamin, is a water-soluble B-vitamin and is synthesized by bacteria. Many meat eaters do not absorb B12 no matter how much meat and animal products they consume. Just as in the case of vitamin A, vitamin B12 deficiency is not endemic among vegans. Dr. Jennifer Rooke writes,

> The Framingham Offspring study found that 39 percent of the general population may be in the low normal and deficient B12 blood level range, and it was not just vegetarians or older people. This study showed no difference in the B12 blood levels of younger and older adults. Most interestingly there was no difference between those [who] ate meat, poultry, or fish and those who did not eat those foods. The people with the highest B12 blood levels were those who were taking B12 supplements and eating B12 fortified cereals.[11]

Furthermore, it must be pointed out that neither animals nor plants manufacture vitamin B12. As I said, Vitamin B12 is the name commonly used to refer to a byproduct of bacteria. This is a very important point that is seldom acknowledged, that is, no animal or plant manufactures B12. As Dr. Rooke points out, "In order to maintain meat a source of B12 the meat industry now adds it to animal feed, 90% of B12 supplements produced in the world are fed to livestock."[12] Consequently if farm animals are given B12 supplements, vegan parents who are concerned about B12 absorption, as well as the morality of killing animals, have good reason to not raise their children on a diet that includes meat and animal products. The most sensible way to avoid B12 deficiency is to get B12 supplements, which are relatively inexpensive and most importantly not derived from animal source. In other words, if animals themselves require B12 in the form of supplements, what is the point of eating meat to get B12? Vegan parents can easily give their children B12 supplements just like farmers give B12 supplements to their animals. No middleman is needed. There are several reasons why B12 is not absorbed. One reason is hypothyroidism.[13] And in general low iodine consumption is the determining factors for the occurrence of hypothyroidism.[14] Another reason is that most people eat cooked food. Cooking simply degrades vitamins. Vitamin B12 in particular is affected by heat. Also, intake of alcohol and coffee can destroy vitamin B12.[15]

With regard to vitamin D, the assertion that vegans run the risk of vitamin D deficiency is inaccurate. There are three points to be considered. First, as in the case of many other micro nutrients, some people for a variety of reasons cannot properly absorb vitamin D. Second, vitamin D synthesis occurs as a result of the sun's ultraviolet rays hitting cholesterol in the skin cells. In normal conditions, the body stores the vitamin for quite a long time.[16] Third, upon review of pertinent literature, it is evident that there is "no association between s25(OH)D concentrations and vegetarian status in either our black or white cohorts. This would indicate that factors other than diet have a greater effect on s25(OH)D than vegetarian status."[17]

Once again, there is scientific evidence that both meat eaters and vegans or vegetarians have lower-than-recommended percentage of vitamin D because most foods contain a lower-than-desirable amount of the vitamin. There is no evidence that vegans in particular are at risk of vitamin D deficiency. Most importantly, there is no evidence that any long-term vegans became ill or died as a result of low vitamin D. Vitamin D is critical for both vegetarian and meat eaters because it is not easy to obtain, except from sun exposure. For this reason, many foods, such as cereal and juices, are fortified. At any rate, any conscientious vegan parents concerned about their children's having desirable levels of vitamin D can easily make sure that their children get adequate sun exposure or sunlamps exposure or use of supplements. Having considered that desired level of vitamin D can be obtained through exposure to sunlight, sunlamps, certain vegetables, fortified foods, and inexpensive supplementation, parents do not need to raise their children on a diet that includes animal products.

Regarding "iodine, calcium, iron, various fatty acids, creatine and taurine" rather than discussing them individually, it is sufficient to make the following four remarks. First, once again, people may be deficient in any of these micronutrients for a variety of reasons irrespective of whether they follow a vegan diet. Consider iodine. Animals do not manufacture iodine, only plants do. Sea vegetables are rich in iodine. Again, why the middleman? Also, there are other fruits and vegetables that contain reasonable amounts of iodine. Furthermore, one may obtain iodine from table salt and supplements. Most importantly, according to the National Institute of Health, the people who may not get enough iodine are "People who do not use iodized salt, pregnant women, people living in regions with iodine-deficient soils who eat mostly local foods, and people who get marginal amounts of iodine and who also eat foods containing goitrogens [such as cruciferous vegetables]."[18] Note that vegans and vegetarians are not included in the list because there is no evidence that only vegans are at risk of not getting enough iodine.

Also, considering all the anti-vegan propaganda and how popular veganism is nowadays, meat eaters cannot wait to prove vegans wrong. The fact is that no case of vegans, adults or children, dying or becoming ill as a result of following vegan diets has yet occurred. The explanation, in my view at any rate, is simple: it is almost impossible not to get the required nutrients from a vegan diet and thrive. One would have to be malnourished, and thus malnourishment would be the cause of health problems and not veganism.

As already mentioned, it is quite easy to meet optimal amounts of these critical micronutrients through a simple vegan diet and, in some cases, supplementation. Hunt seems unnecessarily preoccupied with the notion of appropriately planned diet: "There is a lack of information about the extent to which vegan parents appropriately plan their children's diets."[19] But how complicated is to follow an appropriate vegan diet? Virtually all nutritionists and nutrition experts will recommend the same principle, that is, eat plenty of fresh fruit, vegetables, nuts and seeds, and avoid processed and refined foods.[20] This principle does not apply only to vegans, but also to meat eaters. Any diet lacking in fresh fruit and vegetables will lead to health problems. The addition of meat and animal products will not make the situation better.

Hunt mentions taurine and creatine as potential dangers for vegans. This is a misconception in light of the fact that taurine and creatine are naturally synthesized in the body. Cats need meat because their bodies do not make taurine, but humans do. Taurine is known as a conditionally essential amino acid and creatine is a non-essential amino acid; the body can synthesize both taurine and creatine.[21] Consequently, it is not necessary to eat a meat-based diet to ensure proper levels of taurine and creatine. Therefore, taurine and creatine are in no way potential dangers for vegans.

Finally, as a general remark on raw veganism and children, it must be emphasized that health sciences are now positive that diets rich in fresh, uncooked fruit and greens are beneficial to all age groups and individuals. Researchers constantly show that diets rich in vegetables, fruits, nuts, seeds, herbs, are associated with a lower risk of chronic diseases.[22] Moreover, raw vegan diets can meet satiety needs and energy requirements even to competitive athletes.[23] Consequently, raw vegan diets consisting in abundant servings of fresh, unprocessed and uncooked fruit and greens, nuts and seeds, is the best possible diet for children.[24]

On the other hand, the harm that animal-based diets can cause is quite evident. I will list the most salient examples:

The Cancer Council New South Wales states,

> There is now a clear body of evidence that bowel cancer is more common among those who eat the most red and processed meat. Processed meat consumption has also been strongly linked to a higher risk of stomach cancer.[25]

According to Maryam S. Farvid, a visiting scientist at Harvard School of Public Health,

> One serving a day increment in red meat intake during adolescence was associated with a 22% higher risk of premenopausal breast cancer and each

serving per day increment during early adulthood was associated with a 13% higher risk of breast cancer overall.[26]

According to the Physicians Committee for Responsible Medicine,

> Milk and other dairy products are the top source of saturated fat in the American diet, contributing to heart disease, type 2 diabetes, and Alzheimer's disease. Studies have also linked dairy to an increased risk of breast, ovarian, and prostate cancers.[27]

A study published in *The Journal of Nutrition* concludes that, "this study showed that higher intakes of fish were significantly associated with higher incidence rates of breast cancer. The association was present only for development of ER + breast cancer."[28]

> According to researchers Turner-McGrievy, Mandes, & Crimarco, Both clinical trials and observational research indicate an advantage to adoption of PBDs [Plant-Based Diets] for preventing overweight and obesity and promoting weight loss." and "More than two-thirds (69%) of U.S. adults are overweight.[29]

According to Cross, Koebnick and Sinha,

> Given the plausible epidemiologic evidence for red and processed meat intake in cancer and chronic disease risk, understanding the trends and determinants of meat consumption in the U.S., where meat is consumed at more than three times the global average, should be particularly pertinent to researchers and other public health professionals aiming to reduce the global burden of chronic disease.[30]

According to Orlich, Singh, Sabaté et al.,

> Vegetarian diets are associated with lower all-cause mortality and with some reductions in cause-specific mortality.... These favorable associations should be considered carefully by those offering dietary guidance.... Vegetarian dietary patterns have been associated with reductions in risk for several chronic diseases, such as hypertension, metabolic syndrome, diabetes mellitus, and ischemic heart disease (IHD).[31]

According to a recent study,

> Among US adults, higher consumption of dietary cholesterol or eggs was significantly associated with higher risk of incident CVD [cardiovascular disease] and all-cause mortality in a dose-response manner.

These results should be considered in the development of dietary guidelines and updates.[32]

Furthermore, many Americans fail to reach required micronutrient intake. About 75–80% of the US population do not consume adequate intake of fruit and vegetables.[33]

Finally, as aptly stated by John D. Grant,

> We humans do not need meat. In fact, we are healthier without it, or at least with less of it in our diets. The Adventist Health Studies provide solid evidence that vegan, vegetarian, and low-meat diets are associated with statistically significant increases in quality of life and modest increases in longevity. The world that we inhabit would also be healthier without the commercial meat industry. Factory farms are a waste of resources, environmentally damaging, and ethically indefensible. It is time to accept that a plant-predominant diet is best for us individually, as a race, and as a planet.[34]

In light of the foregoing scientific findings, it is evident that premise 1a of Hunt's argument (Raising children on a vegan diet can harm children physically) is implausible. I want to highlight two reasons. The first is that the studies that Hunt cites are inconclusive for the reason I explained above. Namely, research shows that micronutrient deficiencies are not endemic among vegans due to their diets. In fact, as the studies cited above (Fulgoni, Keast, Bailey & Dwyer, 2011) indicate, many Americans do not reach required micronutrient intake and 75–80% of the US population do not consume enough fruit and vegetables. These percentages are frightening, but what's important to understand is that they do not refer to vegans. The simple explanation is that a meat-based diet does not guarantee proper intake of micronutrients, or optimal health for that matter. Thus, Hunt's assertion that vegan diets in particular bear the risk of harming the health of children is quite implausible.

Consequently, on the grounds of overwhelming scientific evidence showing that consumption of meat and animal products can be unhealthful while plant-based diets are healthful, parents who are committed to whichever ethical principles that led them to embracing veganism have more than good reasons to raise their children on a vegan diet. The second reason is perhaps anecdotal, but nonetheless valid. Hunt seems to overstate his case. If the evidence that he presents suggests that a vegan diet can bare the risk of harming children (or adults for that matter), one would expect to see evidence of children who became ill or died as a result of a not-well-planned vegan diet. The fact is that there is no such evidence. Considering the importance and popularity of this discussion if such evidence were available by now it would have gone viral. On the contrary, as stated above, there is important scientific evidence showing that vegans are healthier than meat eaters.

Social concerns

Next we turn to premise 1b (Raising children on a vegan diet can harm children socially). According to Hunt, "By eating a vegan diet, in particular on those occasions where their peers are eating animal products (a playdate, school lunches, etc.), children are likely to be identified as vegans by their peers and to identify themselves as such."[35] He describes the social interactions that vegans experience in life as "mental whiplash." Namely, the probing and questioning and comments of others are uncomfortable for adults and even more uncomfortable for children. This seems a bit of an exaggeration, especially if it is considered that people follow disparate diets for ethical, health, or religious reasons. The point is that it would seem implausible that ethical vegan parents raised their children on an animal-based diet to avoid the risk of their children being questioned about veganism. People, unfortunately, criticize or make uncomfortable comments about or question other people's religious beliefs, sexual orientation, diet, and more. It does not follow, however, that one should conform in order to avoid uncomfortable comments.

Furthermore, vegan children and non-vegan children alike have all sorts of diets: celiac, kosher, Halal, nuts-and-seeds free, dairy-free, paleo, gluten-free, and the list goes on. There is no particular reason to believe that vegan children would be the target of ridicule and scorn just because they are vegans. At any rate, ethical vegan parents may easily avoid possible issues by instructing their children not to announce that they are vegans and the reason why. If asked, one may answer in several different ways: I do not like/I am allergic to meat/dairy. I do not consume animal products according to my religious beliefs. There is no reason why a child who chooses to avoid eating animal flesh for ethical reasons should identify himself or herself as a vegan. At any rate, in this age of diversity, children may choose to identify themselves as vegans and announce it to their classmates. This may be an opportunity for children and educators to teach others about veganism, animal ethics, and sustainability. Nowadays, school children are exposed to children who wear a hijab or a kippah or a turban, those who have two mothers or two fathers, those who cannot eat dairies or peanuts—the list is endless. Veganism is no "stranger" than anything else. Judging by the enormous interest on veganism, it seems unlikely that children will be retaliated against for being vegans considering the increasing interest in veganism.[36]

The next point, according to Hunt, "In omnivorous societies many cultural festivities (Christmas, Passover, Eid al-Adha, Halloween, birthdays) involve the consumption of food containing animal products." In addition to holidays, people socialize at events that involve food consumption, such as birthday parties and other events. Since vegan children participate in these events, their participation requires certain accommodations that aren't always possible. Here I want to make three points: first, as already mentioned, people have all sorts of diets. It seems unlikely that every single person at a social gathering eats everything that is served. Some people may be diabetic, celiac, kosher, and more, and consequently would require certain accommodations. A Jain child or a Muslim child or a celiac child would require accommodations. Second, as the expression goes, "birds of a feather flock together" and it is likely that vegan parents frequent vegan friends and may tend to participate in social events that are vegan or vegan-friendly. Third, regarding religious celebrations, generally when religious people participate in religious gatherings they are not concerned about food, but rather about (among other things) worshipping God and enjoying each other's company. Christmas, Passover, and other dinner celebrations are typically events organized by families and close friends. While it is true that food is an important aspect of such celebrations, it is hard to imagine that close friends and family members would have any reservations about accommodating children's dietary requirements. Furthermore, it is quite easy for vegan parents to prepare vegan food and bring it to the celebration.

Consider the following scenario: Jason is a vegan child who is invited to his classmate Nashon's birthday party. Nashon's family is all about meat and dairies. At the party, Nashon's parents serve pepperoni pizza and cupcakes. Is Jason's social life ruined? Not at all. Jason's parents can easily arrange vegan pizza and cupcakes to be served—that's assuming that Jason likes pizza and cupcakes and goes to the party in order to eat them. Moreover, most likely, there will be children who will not be able to eat pepperoni pizza and cupcakes for a variety of reasons, children allergic to nuts, celiac children, kosher children, lactose intolerant children, diabetic children, children who require halal food, and the list goes on. Thus, there is no valid reason to believe that the social lives of vegan children in particular will suffer or that vegan children are destined to be social outcast.

Also lacking are reliable studies to determine whether and to what extent vegan children's social lives may be affected by their being vegans. What is available are the testimony of vegan parents and personal experience. These generally do not count high in a scientific study. Nevertheless, they give a sense of whether an argument of the sort proposed by Hunt is valid. These reports, however, confirm my predicted line of argument that a vegan life is not complicated or a daunting experience for children. Emily Moran Barwick is a vegan and an animal rights activist. She has interviewed a number of vegan parents about the social aspects of veganism on the lives of their children. Here are a few examples:

Jesse, The Bronx, NYC:

> Well, for us, like maybe the first one or two times when we went to parties when he was young we didn't know what to do, so it was a little bit hard. So what we've learned to do is we bring our own foods where we go. We pack his lunch with fruits, vegetables, maybe a peanut butter and jelly sandwich.

And a lot of our friends and family have almost—we've turned the tables on them where they're not completely vegan but they'll, like my family, they'll cook vegan meals when we come over. At least have a certain plate or two vegan, and you know, we've rubbed off on a lot of people, so it hasn't been that hard. If we go to a birthday party, we'll bring him a little piece of vegan cake or something just so when they whip out the cake, we can have his piece ready or a cupcake, or something. It's all about preparation. It's not that difficult.

Shantelle, Cambridge, MA:

A lot of times, if you bring vegan cupcakes and people try them and say, "oh I really can't even tell the difference." It'll open up that door for conversation and it can encourage others to try that lifestyle, too.

So, I always try to look at it as a challenge, but as a good challenge because it's an opportunity to educate someone else and show people that this lifestyle–you know–we don't miss anything in this lifestyle. We still have our cookies, we have snacks, we have cakes and ice cream. We have everything that everyone else has except we don't have the guilt.

Theresa:

We always bring our own food, and fortunately we have friends that are very accommodating and they prepare vegan food for us... And I have a few non-vegan friends that introduced me to vegan products that I didn't know existed. But, yeah I try and make something really special like cupcakes or something mouth-watering. Usually the meat eaters eat it all and there's nothing left for us, though.

Kara:

It's quite a good opportunity, actually, taking some really nice vegan treats to show people just how nice the food is that we eat, you know.

Raffaela, Lisbon, Portugal:

It is not difficult at all because I always bring food with me for both of us. And whenever there is a special occasion, like a birthday party, I make sure that he has something that he really enjoys to eat like pretty cupcakes that I make, popcorn, chocolate cookies. He's always happy.

Martin, London, England:

For my youngest son–you always speak to the parents at parties and I always make him aware of what sort of food will be there. So if he goes to a party and they're doing a birthday cake, he knows what ingredients are in there and he doesn't want it because of that. If he did want a cake later in the day we could go home and make him one. And a better one, as well.

Crystal, Virginia:

I feed them first. I make sure they're full. And then, I pack them food. And then, when they go to parties I ask the host if there's anything I can bring. Like, if they have cakes and cupcakes, I will make my own vegan cupcakes and have them bring it so they can eat with the rest of the kids. Fruits and vegetables are not a hassle. You don't even have to cook it, just slice it up, wash it, we can eat it all day. So, that's what we do, especially with AJ. I just pack him some snacks and lunch and make sure he's fed. He doesn't complain.

Ellen, Maui, Hawaii:

So if we start deciding to serve parties, serve food at our own parties that we have, with healthy, delicious food, it can become the norm. We are the first ones that can create change. We have to be the ones, if we want the world to change in any way, if we want it to be normal for our kids to eat delicious, healthy food, we have to be the ones to start it.[37]

What emerges from these testimonies is that it is not complicated at all to raise children on a vegan diet. Most important, vegan children's social lives are not harmed by their being vegans. Parents nowadays are very sensitive about children's dietary restrictions. At birthday parties, it is easy for hosts to prepare vegan food for their guests and for vegan parents to bring vegan food to the parties. As a parent points out, if the child wants a particular food, he or she knows that he or she can have it at home. Hunt is worried that some vegan parents are afraid of being identified as those who always bring their own snacks.[38] I want to make two points: one, as I mentioned a number of times, vegans are not the only ones who bring their own snacks. Children have many different food restrictions due to health conditions, religious, or ethical reasons. Second, it is peculiar to suggest that vegan parents should give up their deepest moral conviction and choose not to raise their children on a vegan diet just to avoid uncomfortable questions or avoid bringing snacks to birthday parties. It is like saying that there is pro tanto moral reason for Jain parents who live in the US not to raise their child as a Jain because Christianity is the prevalent religion and children may be questioned and ridiculed by others. A moral conviction such as ethical veganism should not be bargained with or given up for practical reasons.

Oddly Hunt writes, "a tofurky [sic] seems destined to be a somewhat inferior substitute for a turkey in terms of its cultural and aesthetic cache."[39] This is just a biased assertion. Also, it evinces a naïve understanding of veganism. Not all vegans eat or like Tofurkey. For example, neither my children who have been vegans since they were born nor I eat Tofurkey. Children who choose to be vegans understand the moral reasons for veganism.[40] Many people are in fact vegans because they cannot stand even the idea of a dead bird on a table. Thus, a vegan child may actually regard the real turkey as inferior to Tofurkey. Also, Hunt's argument exaggerates the importance of food and assumes that children (and adults) participate in events just to eat. However, religious and social events are meant as opportunities for friends and families to be together and enjoy each other's company. Similarly, children don't go to events with the specific goal to eat, but rather to spend time with their friends and play.

On the equal strength of 4) and moral reasons for veganism

Lastly, Hunt's argument extends to 4): "This pro tanto moral reason is plausibly as strong as the pro tanto moral reason some vegan parents have, given their preferred moral framework, to raise their child on a vegan diet" and to conclusion 5): "that these parents may plausibly find it morally permissible to raise their child on a non-vegan diet."[41] Hunt is quite right that ultimately the question of raising children on a vegan diet or not rests on the fact that "Many vegans are non-philosophers. This means that in many cases they have no elaborate philosophical framework, such as Regan's, from which they derive moral reasons for being vegan."[42] The problem is that 4) is not as plausible as the moral reasons that vegan parents have to raise their children on a vegan diet. Thus, 3) and 4) fail to support the conclusion 5), that parents have plausible reasons to find it morally permissible to not raise their children on a vegan diet. Even though most vegan parents are not philosophers with sophisticated moral theories in their arsenals, they have facts that generate moral reasons against feeding their children animal products.

Hunt suggests that non-philosopher vegan parents might rely upon the following principles:

R1 "Do not harm or kill animals, except when necessary."

R2 "Do not harm or kill animals, unless refraining from doing so causes you or yours a significant harm."

R3 "Do not harm or kill animals for the sake of trivial pleasures, such as the gustatory."

R4 "Do not harm or kill animals, unless they pose some harm to you."[43]

However, according to Hunt "many such principles, such as R1–3, do not seem to generate moral reasons against feeding one's child animal products."[44] Consider the following principles upon which non-philosopher vegan parents might rest their choice to be vegans:

1   Intensive animal farming causes a great deal of unnecessary suffering to animals.[45] Furthermore, intensive animal agriculture involve unnecessarily violent practices. We do not wish to support those practices, and consequently we do not want our children to support those practices. We refuse to contribute to animal suffering; and consequently refuse to raise our children on an animal-based diet for the same reason. Therefore, we have a moral obligation to raise our children on a vegan diet.
2   Animal agriculture is the leading cause of environmental degradation.[46] We have a moral responsibility toward the environment to avoid environmental degradation so that future generations will have a healthy environment. A vegan diet is environmentally friendly, while animal-based diets have negative effect for the environment.[47] Consequently, we have a moral obligation to raise our children on a vegan diet.
3   Scientific studies continue to show a significant link between animal-based food and various diseases, such as cancer, diabetes, high blood pressure, heart disease, and more. Also, they show that vegans and vegetarians are healthier and may live longer than people who consume animal products.[48] We find it easier to plan a healthful vegan diet than to risk that our children will suffer from diseases caused by an animal-based diet. Thus, we have a moral obligation to raise our children on a diet that does not bare the risk of chronic diseases.

The foregoing principles may not be philosophically rigorous, but they clearly generate strong moral reasons against raising children on an animal-based diet. These principles are very basic. Any layperson with a basic education understands the gravity of the issues that principles 1) – 3) raise. Animal suffering, environmental degradation, and chronic diseases, can be avoided by eating a vegan diet. Thus, considering Hunt's second premise ("Parents have pro tanto moral reason to not make their child bear a risk of harm to their well-being"), it is more plausible for vegan parents to raise their children on a vegan diet because such a diet does not bear a risk to harm their children's well-being as well as the animals' and the environment's.

Consider also the following comments of vegan parents:

Mandi Roberts:

> It wasn't until I was around 7, that I realized what animals actually went through. I remember talking about it as a child, but it was quickly swept under the carpet.

Rebecca Coplon:

> My son was about 3 when he got the idea that chicken was, well, made of chicken, but about 4 when he put two and two together and realized he was EATING A CHICKEN! He's nearly 12 now. Even in Minecraft, he is a vegetarian, and resists killing the little digital beasts to get meat or leather.

Erica Challis:

> At around 3, for our son. He stopped eating meat because it was made of animals, he said. At six, he still hasn't eaten meat since he was three.

Subha Thankaraj:

> I was showing my 3 and half year old daughter the video of dairy product factory. She asked, "So, now do they give this milk to baby cow?" I replied, "No. It's transported to supermarkets and we buy it." Then, she asked "So, what do baby cows drink?"

Germaine Cecil:

> When I was 7 I realized that animals had to be killed for me to eat … and the conditions of being force fed and so forth I stopped eating meat…. I've been vegetarian ever since.[49]

The foregoing comments are anecdotal, but nonetheless show something important, that is, children naturally understand that animals are not food. It also suggests that in most cases discussions about the origins of meat and animal products are avoided or played down by caregivers.[50] When children become aware of the origin of animal products, when they connect the dots, as it were, they feel empathy for the animals and choose not to eat them.[51] The point is that children's moral awareness about the unjust treatment of farm animals gives vegan parents another good reason to raise their children on a vegan diet. As we have seen, one of the flaws in Hunt's argument is that premise 1), which consists in 1a and 1b, is false. Consequently, they fail to support 3), that parents have good reasons to not raise their children on a vegan diet. Premise 4), as we have seen is not true, that is, it is not plausible that the moral reasons for raising children on a vegan diet are equally strong as the moral reasons that vegan parents have to raise their children on an animal-based diet. Therefore, the conclusion, premise 5), that vegan parents may plausibly find it morally permissible to raise their child on an animal-based diet does not follow.

Conclusion

Therefore, I have shown that vegan parents have valid reasons, on the basis of very practical moral frameworks, to conclude that it is *not* morally permissible to raise their child on an animal-based diet. Vegan parents that are concerned with their children's well-being, have a moral obligation to raise them on a vegan diet because animal-based diets can lead to many diseases, optimal vegan diets do not require complicated planning, and a vegan diet in no way harms a child's social life. In fact, raw vegan diets are the most optimal because of their focus on highly nutritious, raw fruit, greens, nuts, and seeds. Most children who go vegan because they are morally against animal suffering and exploitation and, consequently, not eating a pepperoni pizza or a dairy-based cake at a party is not a problem for at least two reasons: one, those foods are made from the exploitation of animals. Thus, vegan children do not desire to eat them. Two, vegan children do not miss anything because they know that if they have the desire for those foods, they can have vegan pepperoni pizza and vegan cake. Also, vegan diets do not cause as much animal suffering as animal-based diets, and vegan diets are environmentally friendlier that meat-based diets, raw vegan diets being the most environmentally friendly. Furthermore, when children are informed about the origin of animal products, they show empathy for animals and choose not to eat them. Consequently, vegan parents who choose to not raise their children on a vegan diet would have to betray not only their moral commitments, but also potentially harm their children's health and ethical consideration regarding animals.

Notes

1 Marcus William Hunt, "Veganism and children: Physical and social well-being." *Journal of Agricultural and Environmental Ethics*, 32, 269, 2019. https://doi.org/10.1007/s10806-019-09773-4. 269.
2 Hunt, 2019: 269.
3 W. J. Craig and A. R. Mangels, "Position of the American Dietetic Association: Vegetarian diets." *American Dietetic Association*, 109(7), 2009: 1266–1282; Vesanto Melina, Winston Craig, Susan Levin et al. "Position of the academy of nutrition and dietetics: vegetarian diets." *Journal of the Academy of Nutrition and Dietetics*, 116(12), 2016: 1970–1980.
4 Sílvia Castañé and Antón Assumpció, "Assessment of the nutritional quality and environmental impact of two food diets: A Mediterranean and a vegan diet." *Journal of Cleaner Production*, 167(1), 2017: 929–937. doi:10.1016/j.jclepro.2017.04.121
5 Office of Disease Prevention and Health Promotion (ODPHP). *Dietary Guidelines 2015–2020*, n.d. https://health.gov/dietaryguidelines/2015/guidelines/chapter-2/current-eating-patterns-in-the-united-states/: para. 1.
6 ODPHP, n.d.: para. 1.
7 Hunt, 2019: 271.
8 Ibid.
9 J. McBride, (2000). "B12 deficiency may be more widespread than thought." United States Department of Agriculture Agric. Res. Service, 2000. www.ars.usda.gov/news-events/news/research-news/2000/b12-deficiency-may-be-more-widespread-than-thought/
10 J. Higdon, Linus Pauling Institute, Micronutrient Information Center. (2000). https://lpi.oregonstate.edu/mic/vitamins/vitamin-A
11 Jennifer Rooke, "Do carnivores need Vitamin B12 supplements?" *Baltimore Post-Examiner*, 2013. https://baltimorepostexaminer.com/carnivores-need-vitamin-b12-supplements/2013/10/30
12 Rooke, 2013, para. 4.
13 A. Jabbar, A. Yawar, S. Waseem, S. et al. "Vitamin B12 deficiency common in primary hypothyroidism." [Published correction appears in *J. Pak. Med. Assoc.*, 59 (2), 2009: 126]. *J. Pak. Med. Assoc.*, 58(5), 2008: 258–261.
14 M. P. Vanderpump and W. M. Turnbridge, "Epidemiology and prevention of clinical and subclinical hypothyroidism." *Thyroid*, 12, 2002: 839–847.

15 A. Ulvik, S. E. Vollset, G. Hoff and P. M. Ueland, "Coffee consumption and circulating b-vitamins in healthy middle-aged men and women." *Clin. Chem.*, 54, 2008: 1489–1496.
16 M. Wacker and F. M. Holick, "Sunlight and Vitamin D." *Dermato-Endocrinology*, 5 (1), 2013: 51–108. doi:10.4161/derm.24494
17 J. Chan, K. Jaceldo-Siegl and E. G. Fraser, "Serum 25-hydroxyvitamin D status of vegetarians, partial vegetarians, and nonvegetarians: The Adventist Health Study-2." *The American Journal of Clinical Nutrition*, 89(5), 2009: 1686S–1692S. doi:10.3945/ajcn.2009.26736X
18 National Institute of Health. "Iodine." 2016. https://ods.od.nih.gov/factsheets/Iodine-Consumer/#h5
19 Hunt, 2019: 271.
20 Mayo Clinic. Patient Care and Health Info. Nutrition and Healthy Eating, 2018. www.mayoclinic.org/healthy-lifestyle/nutrition-and-healthy-eating/in-depth/vegetarian-diet/art-20046446
21 R. Lourenco and M. E. Camilo, "Taurine: a conditionally essential amino acid in humans? An overview in health and disease." *Nutricion hospitalaria*, 17(6), 2002: 262–270.; Greenhaff, L. P. "The nutritional biochemistry of creatine." *The Journal of Nutritional Biochemistry*, 8(11), 1997. doi:10.1016/S0955-2863(97)00116–00112
22 Julieanna Hever and Raymond J. Cronise. "Plant-based nutrition for healthcare professionals: implementing diet as a primary modality in the prevention and treatment of chronic disease." *Journal of Geriatric Cardiology: JGC* vol. 14,5 (2017): 355–368. doi:10.11909/j.issn.1671–5411.2017.05.012.
23 Roman Leischik and Norman Spelsberg, "Vegan Triple-Ironman (raw vegetables/fruits)." *Case Reports in Cardiology*, 2014, 2014: 317246. doi:10.1155/2014/317246
24 Luciana Baroni et al. "Vegan nutrition for mothers and children: Practical tools for healthcare providers." *Nutrients*, 11(1), 2018: 5. doi:10.3390/nu11010005
25 Cancer Council NSW (n.d.). *Meat and Cancer*. www.cancercouncil.com.au/21639/cancer-prevention/diet-exercise/nutrition-diet/fruit-vegetables/meat-and-cancer/
26 S. M. Favid, "Red meat consumption and breast cancer risk." *Harvard School of Public Health*, 2014. www.hsph.harvard.edu/news/features/red-meat-consumption-and-breast-cancer-risk/
27 Physicians Committee for Responsible Medicine (n.d.). "Health concerns about dairy: Avoid the dangers of dairy with a plant-based diet." www.pcrm.org/good-nutrition/nutrition-information/health-concerns-about-dairy
28 C. Stripp, K. Overvad, J. Christensen, B. L. Thomsen, A. Olsen, S. Møller et al. "Fish intake is positively associated with breast cancer incidence rate." *The Journal of Nutrition*, 133(11), 2003: 3664–3669. doi:10.1093/jn/133.11.3664
29 G. Turner-McGrievy, T. Mandes and A. Crimarco, "A plant-based diet for overweight and obesity prevention and treatment." *Journal of Geriatric Cardiology*, 14(5), 2017: 369–374. doi:10.11909/j.issn.1671-5411.2017.05.002: 369.
30 C. Daniel, A. Cross, C. Koebnick and R. Sinha, "Trends in meat consumption in the USA." *Public Health Nutrition*, 14(4), 2011: 575–583. doi:10.1017/S1368980010 002077: 1.
31 M. J. Orlich, P. N. Singh, J. Sabaté, K. Jaceldo-Siegl, J. Fan, S. Knutsen and G. E. Fraser, "Vegetarian dietary patterns and mortality in Adventist Health Study 2." *JAMA Internal Medicine*, 173(13), 2013: 1230–1238. doi:10.1001/jamainternmed.2013.6473
32 V.W. Zhong, L. Van Horn, M. C. Cornelis, J. T. Wilkins, H. Ning, M. R. Carnethon, P. Greenland, R. J. Mentz, K. L. Tucker, L. Zhao, A. F. Norwood, D. M. Lloyd-Jones and N. B. Allen, "associations of dietary cholesterol or egg consumption with incident cardiovascular disease and mortality." *JAMA*, 19(11), 2019: 1081–1095. doi:10.1001/jama.2019.1572

33 V. L. Fulgoni, D. R. Keast, R. L. Bailey and J. Dwyer, "Foods, fortificants, and supplements: Where do Americans get their nutrients?" *The Journal of Nutrition*, 141 (10), 2011: 1847–1854. doi:10.3945/jn.111.142257
34 J. D. Grant, "Time for change: Benefits of a plant-based diet." *Canadian Family Physician Medecin de famille canadien*, 63(10), 2017: 744–746.
35 Hunt, 2019: 275.
36 D. Hancox, "The unstoppable rise of veganism: how a fringe movement went mainstream." *The Guardian*, 2018. www.theguardian.com/lifeandstyle/2018/apr/01/vegans-are-coming-millennials-health-climate-change-animal-welfare
37 Moran E. Barwick, "Bite size vegan. Are vegan kids social outcasts? Parents answer." 2016. www.bitesizevegan.org/bite-size-vegan-nuggets/qa/are-vegan-kids-social-outcasts-parents-answer/
38 Hunt, 2019: 277.
39 Ibid.
40 K. M. Hussar and P. L. Harris, "Children who choose not to eat meat: A study of early moral decision-making." *Social Development*, 19, 2010: 627–641. doi:10.1111/j.1467-9507.2009.00547.x
41 Hunt, 2019: 281 & 287.
42 Ibid.: 286.
43 Ibid.: 286.
44 Ibid.: 287.
45 D. Wicks, "Silence and denial in everyday life: The case of animal suffering." *Animals: An Open Access Journal from MDPI*, 1(1), 2011: 186–199. doi:10.3390/ani1010186
46 J. Cameron and A. S. Cameron, "Animal agriculture is choking the Earth and making us sick. We must act now." *The Guardian*, 2017. www.theguardian.com/commentisfree/2017/dec/04/animal-agriculture-choking-earth-making-sick-climate-food-environmental-impact-james-cameron-suzy-amis-cameron; G. Koneswaran and D. Nierenberg, "Global farm animal production and global warming: Impacting and mitigating climate change." *Environmental Health Perspectives*, 116(5), 2008: 578–582. doi:10.1289/ehp.11034
47 Tuomisto, L. H. Importance of considering environmental sustainability in dietary guidelines. The Lancet Planetary Health. (2018). https://doi.org/10.1016/S2542-5196(18)30174-8.
48 S. M. Alshahrani, Gary E. Fraser, J. Sabaté, R. Knutsen, D. Shavlik, A. Mashchak, J. I. Lloren and Michael J. Orlich, "Red and processed meat and mortality in a low meat intake population." *Nutrients* 11(3), 2019: 622. doi:10.3390/nu11030622; Stripp et al., 2003; Zhong et. al., 2019.
49 "When do kids realize that eating meat involves killing animals?" *Quora*. www.quora.com/When-do-kids-realize-that-eating-meat-involves-killing-animals
50 Brian Luke, "Justice, caring, and animal liberation." *Between the Species*, 8(2), 1992, Article 13.
51 Josephine Donovan, "Feminism and the treatment of animals: From care to dialogue." *Signs*, 31(2), 2006: 305–329.

8 Conclusion

Why do people eat animal products? It is a complicated question, just like many other aspects of life, traditions, beliefs, and customs. The short answer is that virtually all societies in the world are imposed upon animal-based food. As a result, many people have come to believe that consuming animal-based diets is essential. However, we know that consuming animal products is not essential to good health. In fact, nutrition science shows that animal products can be very unhealthful to humans, and should be consumed in moderation. Moreover, using animals for food has proven to be a leading cause of environmental degradation. Many people argue that animal products are tasty. While I do not intend to dispute the subjective, positive experience of eating animal products, I want to make two points: First, while enjoying the taste of food is an important aspect of a good life, the gustatory value of eating animal products is unimportant when its cost is factored; and its cost is environmental degradation and animal exploitation. In addition to their negative environmental impact, the practices involved in producing animal-based food are less than noble; they exemplify the worst vices of which human beings are capable, such as callousness, gratuitous violence, injustice, and intemperance. Second, animal products are not inherently tasty. It is not eating animal flesh that people really enjoy, but rather the tradition and flavors that are added to animal products by cooking them. Meat are typically prepared with spices, marinated, cured, smoked, and cooked because, after all, it is rotting flesh. In fact, as I mentioned previously (Chapter 5) raw meat is hard to chew and uneasy to digest. Consequently, as I argued in Chapter 2, eating animal products is a form of self-deception. Since self-deception should be avoided, it follows that one should not consume animal products.

Another reason people eat animal products is that it is believed to be "natural." According to this view, human beings are somehow "designed" by nature or by God to eat animals, and thus eating meat is the cycle of life. Considering that most people in affluent societies buy conveniently packaged animal products in supermarkets, it is not clear in what sense this represents "the cycle of life." Perhaps, it is more appropriate to say that it is the cycle of supermarkets. Also, it is quite evident from human physiology and anatomy that human beings are not animal eaters. I just cannot imagine how anyone who has access to an abundance of fresh fruit and greens could consider cows, pigs, chickens, lambs, cats, dogs, and other animals in their natural environment, grunting, scurrying, grazing, and find them appetizing. Granted, during periods of scarcity eating animals and insects helped humans survive. But one thing is survival (starving) food, and quite another is what is optimal to the human body and makes us thrive. Therefore, eating animal products is not, in the sense indicated above, natural.

What is natural for humans is eating a raw vegan diet. What I mean by raw may be different for another person. The best possible approach to raw veganism is to consume an abundance of fresh fruit and tender leafy greens, excluding foods and beverages that are unnecessary to good health, such as tea, coffee, alcohol, refined sugars, and salt.

Humans have been around for about 200,000 years (not to mention that hominins have been around for millions of years). During this time, early hominins would eat fruit and tender leafy greens. These are the most nutritiously dense foods, and the most beneficial source of energy for humans. No significant change occurred to our digestive system that equipped us for digesting cooked food. Thus, considering that human beings are evolved creatures, adapted to their environment, it follows that there is a diet that is specific and optimal for our species. Cooking food is a relatively new practice for humans. For the longest time, humans have eaten fruit and tender leafy greens. This is a scientifically documented fact—humans are frugivores. Consequently, cooking food is in no way beneficial to human health. The only benefit is that it provides easy calories by heat-processing food that otherwise would be indigestible. All other species in nature have their own specific diet required to maintain optimal health. Unless one argues that human beings are an exception, we must conclude that humans also have a specific diet to which they are physiologically adapted in order to live healthfully. Humans are frugivores, which is evinced by their being physiologically equipped to obtain energy primarily from the sugar in fruits. Our anatomy enables us to pick fruits, masticate, and digest, with ease and efficiency.

The human diet is rich of all necessary micronutrients. On the other hand, cooking changes the molecular structure of food—and not for the better. It destroys nutrients, creates acrylamides and other carcinogenic substances, and denatures proteins, which leads to many problems. Just to mention one problem: Leukocytosis is an increase of white blood cells in the body. This occurs as a reaction to inflammations or infections. In other words, when the body detects a threat, as a response, it produces more white blood cells. Leukocytosis, however, does not occur when we eat food in its raw state. However, it does occur whenever we consume any type of cooked food. Furthermore, cooking food—vegan or not—shortens our lives[1] while consumption of fruit and greens extends it.[2] In the words of urologist Donald S. Coffey,

> In summary, we were not biologically selected by the evolution process to eat the way we do today, and the damage is manifested in prostate and

breast cancer. Indeed, all of the present suggestions of the National Cancer Institute and the American Cancer Society as to how Americans might reduce their chances of getting prostate and breast cancer revolve around adapting dietary changes in our lifestyle back toward the early human diet of more fruits; a variety of fresh vegetables and fiber; less burning, cooking, and processing; diminished intake of dairy products, red meat and animal fats.[3]

The point of this book is to offer a new way of thinking about food and ethics. As a concluding remark, I would like to highlight some of the benefits of adopting the human diet. The first is that the human diet is our natural diet. The fact that human beings began cooking food is accidental, and thus represents a different question from that of what is our natural and optimal food. Our anatomy and physiology show that we are frugivores. Science confirms this showing that fruit and green vegetables (but particularly fruit) are beneficial in many ways, extend our life, avoid chronic diseases, and more.[4]

Second, the human diet is the best solution to environmental issues. Currently, Livestock systems occupy about 40% of the planet's arable land.[5] Considering that human population in 2050 is estimated to be 10 billion, population growth is an impediment to achieving improvements in food security and species diversity.[6] If the human diet were implemented, the arable land that is currently occupied by the livestock sector would be released, promoting more abundant and diverse harvests to enlarge the human food supply. In other words, it is a known fact that vegan diets are more efficient than animal-based diets. The human diet would be the most efficient.

Third, the human diet would drastically reduce the use of energy resources based on fossil fuels, which are non-renewable, as well as other natural resources.[7] Adopting a raw vegan diet would diminish our dependence on electricity, fossil fuels, water, antibiotics (typically used to keep farm animals' infections under control), and fertilizers. In fact, implementing the human diet could possibly eliminate polluting household waste while producing natural fertilizer.

Fourth, the notion of using animals for food and other purposes represents one of the world's greatest injustices. Failing to treat animals and the environment with respect are certainly less than fully virtuous. Using animals for out benefits and consuming their bodies and their secretions and eggs exemplify a variety of vices, in particular, indifference to the value of nature, ignorance, self-importance, self-indulgence, unfairness, and intemperance. It also evinces a lack of humility, and a sense of beauty. In short, animal exploitation is a failure at human excellence. Conversely, the human diet is the expression of virtue; and the diet that is most harmonious with nature.

I would like to make a final remark by making an analogy between food and a particular event in the 1997 film *Contact* by Robert Zemeckis.[8] The protagonist of *Contact* is Ellie Arroway, played by Jodie Foster. Ellie is a scientist who works for the SETI Institute, hoping to receive some signal from an extra-terrestrial source. One day she detects a radio message from the star system Vega about 26 light-years away. The message contains the blueprints for what seems to be a single-occupant spacecraft for intergalactic travel. Naturally, Ellie is determined to travel wherever the spacecraft takes her. Once the spacecraft is built, the engineers realize that the device is missing a chair for the occupant. Considering that the machine has to be dropped into four rapidly spinning rings, which cause the machine to travel through a series of wormholes, a special chair with seatbelts is added for comfort and safety. It is odd that the designers did not include a seat for the occupant. An oversight perhaps? Once the machine is ready, Ellie is seated and buckled up, and the machine is dropped into the spinning rings. As expected, the ride is quite rough and Ellie experiences a strain from the machine's tremendous speed. The machine is shuddering and rattling like a building in an earthquake. Then Ellie has the right intuition to unbuckle herself before her chair is violently ripped off the floor from the incessant vibrations. At this point she is standing in the middle of the machine and the shuddering and rattling have completely ceased. At this point, Ellie, and the viewer, realizes the omission of a seat was not a mistake. In fact, the mistake was to build a seat. In my view, the current diets of cooked food that humans follow (animal-based or vegan) is like the seat in Ellie's machine. The sooner we "unbuckle" ourselves, the sooner we will achieve physical and moral health.

Notes

1 Sara Arganda et al., "Parsing the life-shortening effects of dietary protein: Effects of individual amino acids." *Proceedings. Biological Sciences*, 284(1846), 2017: 20162052. doi:10.1098/rspb.2016.2052
2 Zumin Shi et al., "Food habits, lifestyle factors and mortality among oldest old Chinese: The Chinese Longitudinal Healthy Longevity Survey (CLHLS)." *Nutrients*, 7 (9), 2015: 7562–7579. doi:10.3390/nu7095353
3 Donald S. Coffey, "Similarities of prostate and breast cancer: Evolution, diet, and estrogens." *Urology*, 57, 2001: 31–38.
4 Heidi Lynch et al., "Plant-based diets: considerations for environmental impact, protein quality, and exercise performance." *Nutrients*, 10(12), 2018: 1841. doi:10.3390/nu1012184
5 H. Steinfeld, P. Gerber, T. Wassenaar, V. Castel, M. Rosales, M. and C. de Haan, "Livestock's long shadow: Environmental issues and options." Rome: FAO, 2006. www.faostat.fao.org.
6 M. Herrero et al., "Livestock and the environment: What have we learned in the past decade?" *Annu. Rev. Environ. Resour.*, 40, 2015: 177–202.
7 Keelia O'Malley et al., "Vegan vs Paleo: Carbon footprints and diet quality of 5 popular eating patterns as reported by US consumers." *Current Developments in Nutrition*, 3(1nzz047), 2019. doi:10.1093/cdn/nzz047.P03-007-19
8 Zemeckis, Robert (dir.), *Contact*. Burbank, CA: Warner Home Video, 1997. DVD.

References

Adams, J. Carol. *The Sexual Politics of Meat: A Feminist-Vegetarian Critical Theory*. Bloomsbury Academic, 2015.

Aiello, C. Leslie and Wheeler, Peter. "The expensive-tissue hypothesis: The brain and the digestive system in human and primate evolution." *Current Anthropology*, 36(2), 1995: 199–221.

Albert, M. J., Mathan, V. I. and Baker, S. J. "Vitamin B12 synthesis by human small intestinal bacteria." *Nature* 283, 1980: 781–782.

Alberts, B., Johnson, A., Lewis, J. et al. *Molecular Biology of the Cell*, 4th edn. New York: Garland Science, 2002. www.ncbi.nlm.nih.gov/books/NBK21054/

Alvaro, Carlo. "Ethical veganism, virtue, and greatness of the soul." *J. Agric. Environ. Ethics* 30, 2017: 765. doi:doi:10.1007/s10806-017-9698-z

Alvaro, Carlo. "Veganism as a virtue: How compassion and fairness show us what is virtuous about veganism." *Future of Food: Journal of Food, Agriculture and Society*, 5(2), 2017.

Alvaro, Carlo. *Ethical Veganism, Virtue Ethics, and the Great Soul*. Lanham: Lexington Books, 2019.

Aleksandrowicz, L., Green, R., Joy, E. J. M., Smith, P. and Haines, A. "The impacts of dietary change on greenhouse gas emissions, land use, water use, and health: A systematic review." *PLoS ONE* 11(11), 2016: e0165797. doi:doi:10.1371/journal.pone.0165797; www.who.int/nutrition/publications/public_health_nut6.pdf

Alshahrani, S. M., Fraser, Gary E., Sabaté, J., Knutsen, R., Shavlik, D., Mashchak, A., Lloren, J. I. and Orlich, Michael J. "Red and processed meat and mortality in a low meat intake population." *Nutrients* 11(3), 2019: 622. doi:doi:10.3390/nu11030622

Alzola, Miguel. "The possibility of virtue." *Business Ethics Quarterly*, 22(2), 2012: 377–404.

American Heart Association. "Nearly half of all adult Americans have cardiovascular disease." *ScienceDaily*. www.sciencedaily.com/releases/2019/01/190131084238.htm (accessed September 20, 2019).

"Amino acids." U.S. National Library of Medicine*MedlinePlus* (n.d.). https://medlineplus.gov/ency/article/002222.htm (accessed August 22, 2019).

Anderson, E. and Barrett, L. "Affective beliefs influence the experience of eating meat." *PLoS ONE*, 11(8), 2016: e0160424. doi:doi:10.1371/journal. pone.0160424www.unep.fr/shared/publications/pdf/dtix1262xpa-priorityproductsandmaterials_report.pdf

Annas, Julia. "Virtue ethics and the charge of egoism." In P. Bloomfield (ed.) *Morality and Self-Interest*. New York: Oxford University Press, 2007.

Animal Kill Clock. https://animalclock.org (accessed June 12, 2018).
Arganda, Sara et al. "Parsing the life-shortening effects of dietary protein: Effects of individual amino acids." *Proceedings. Biological Sciences*, 284(1846), 2017: 20162052. doi:doi:10.1098/rspb.2016.2052
Aristotle. *Nicomachean Ethics*, Sarah Broadie and Christopher Rowe (eds). Oxford University Press, 2002.
Aquinas, Thomas. *Summa Contra Gentiles*, A. Pegis, trans. University of Notre Dame Press, 2016.
Bailey, R. L., Fulgoni, V. L., Keast, D. R. and Dwyer, J. T. "Examination of vitamin intakes among US adults by dietary supplement use." *Journal of the Academy of Nutrition and Dietetics*, 112(5), 2012: 657–663.e4. doi:doi:10.1016/j.jand.2012. 01. 026
Baroni, Luciana et al. "Vegan nutrition for mothers and children: Practical tools for healthcare providers." *Nutrients*, 11(1), 2018: 5. doi:doi:10.3390/nu11010005
Barwick, Moran E. "Bite size vegan. Are vegan kids social outcasts? Parents answer." 2016. www.bitesizevegan.org/bite-size-vegan-nuggets/qa/are-vegan-kids-social-outcasts-parents-answer/
Bhat, Z., Kumar, S. and Fayaz, H. "In vitro meat production: Challenges and benefits over conventional meat production." *Journal of Integrative Agriculture*, 14, 2014: 241. doi:doi:10.1016/s2095-3119(14)60887-x
Birt, D. F., Boylston, T., Hendrich, S., Jane, J.-L., Hollis, J., Li, L., McClelland, J., Moore, S., Phillips, G. J., Rowling, M., Schalinske, K., Scott, M. P., and Whitley, E. M. "Resistant starch: Promise for improving human health." *Advances in Nutrition*, 4 (6), 2013: 587–601. doi:doi:10.3945/an.113.004325
Blocker, J., et al. *Alcohol and Temperance in Modern History*, Vol. 1. Santa Barbara, CA: ABC-CLIO, 2003.
Bouvard, V., Loomis, D., Guyton, K. Z., et al. "Carcinogenicity of consumption of red and processed meat." *The Lancet Oncology*, 16, 2015: 1599. doi:doi:10.1016/s1470-2045(15)00444-00441
Brookie, K. L., Best, G. I. and Conner, T. S. "Intake of raw fruits and vegetables is associated with better mental health than intake of processed fruits and vegetables." *Front. Psychol.*, 9, 2018: 487.
Cameron, J. and Cameron, A. S. "Animal agriculture is choking the Earth and making us sick. We must act now." *The Guardian*, 2017. www.theguardian.com/commentisfree/2017/dec/04/animal-agriculture-choking-earth-making-sick-climate-food-environmental-impact-james-cameron-suzy-amis-cameron (accessed April 19, 2019).
Cancer Council NSW (n.d.). *Meat and Cancer*. www.cancercouncil.com.au/21639/cancer-prevention/diet-exercise/nutrition-diet/fruit-vegetables/meat-and-cancer/
Carmody, R. N. and Wrangham, R. W. "The energetic significance of cooking." *J. Hum. Evol.*, 57, 2009: 379–391. doi:doi:10.1016/j.jhevol.2009.02.011
Carmody, R. N., Weintraub, G. S. and Wrangham, R. W. "From the cover: Energetic consequences of thermal and nonthermal food processing." *Proc. Natl. Acad. Sci. U.S. A.*, 108, 2011: 19199–19203. doi:doi:10.1073/pnas.1112128108
Carus, Felicity. "UN urges global move to meat and dairy-free diet." *The Guardian*, 2010. www.theguardian.com/environment/2010/jun/02/un-report-meat-free-diet (accessed August 18, 2019).
Castañé, Sílvia and Assumpció, Antón. "Assessment of the nutritional quality and environmental impact of two food diets: A Mediterranean and a vegan diet." *Journal of Cleaner Production*, 167(1), 2017: 929–937. doi:doi:10.1016/j.jclepro.2017.04.121

Chan, J., Jaceldo-Siegl, K. and Fraser, E. G. "Serum 25-hydroxyvitamin D status of vegetarians, partial vegetarians, and nonvegetarians: The Adventist Health Study-2." *The American Journal of Clinical Nutrition*, 89(5), 2009: 1686S-1692S. doi:doi:10.3945/ajcn.2009.26736X

Chapman, H. A. and Anderson, A.K.. "Trait physical disgust is related to moral judgments outside of the purity domain." *Emotion*, 14(2), 2014: 341–348.

Chen, Nelson G. et al. "Transient model of thermal deactivation of enzymes." *Biochimica et biophysica acta*, 1814(10), 2011: 1318–1324. doi:doi:10.1016/j.bbapap.2011. 06. 010

Center for Biological Diversity. "How eating meat hurts wildlife and the planet." *Take Extinction off Your Plate* (n.d.). www.takeextinctionoffyourplate.com/meat_and_wild life.html (accessed August 24, 2019).

Centers for Disease Control and Prevention (CDC). "Calorie consumption on the rise in United States, particularly among women." www.cdc.gov/nchs/pressroom/04news/calorie.htm,2014 (accessed June 12, 2019).

Coffey, Donald S. "Similarities of prostate and breast cancer: Evolution, diet, and estrogens." *Urology*, 57, 2001: 31–38.

Cohen, Carl. "The case for the use of animals in biomedical research." *The New England Journal of Medicine*, 315, 1986: 865.

Collard, Andree and Contrucci, Joyce. *Rape of the Wild: Man's Violence Against Animals and the Earth*. Bloomington: Indiana University Press, 1989.

Collins, Nick. "Test tube hamburgers to be served this year." *The Telegraph*, February 19, 2012. www.telegraph.co.uk/news/science/science-news/9091628/Test-tube-hamburgers-to-be-served-this-year.html (accessed August 12, 2019).

Comte-Sponville, A. *A Small Treatise on the Great Virtues: The Uses of Philosophy in Everyday Life*. Metropolitan Books, 2001.

Cornélio, A. M., de Bittencourt-Navarrete, R. E., de Bittencourt Brum, R., Queiroz, C. M. and Costa, M. R. "Human brain expansion during evolution is independent of fire control and cooking." *Frontiers in Neuroscience*, 10, 2016: 167. doi:doi:10.3389/fnins.2016.00167

Craig, W. J. and Mangels, A. R. "Position of the American Dietetic Association: Vegetarian diets." *American Dietetic Association*, 109(7), 2009: 1266–1282.

Cuenca-Sánchez, Marta et al. "Controversies surrounding high-protein diet intake: satiating effect and kidney and bone health." *Advances in Nutrition*, 6(3), 2015: 260–266. doi:doi:10.3945/an.114.007716

Cummings, K. M. and Proctor, R. N. "The changing public image of smoking in the united states: 1964–2014." *Cancer Epidemiology, Biomarkers & Prevention: A Publication of the American Association for Cancer Research*, 23(1), 2014: 32–36. doi:doi:10.1158/1055-9965.EPI-13-0798

Current Eating Patterns in the United States. *Dietary Guidelines 2015–2020*. https://hea lth.gov/dietaryguidelines/2015/guidelines/chapter-2/current-eating-pattern s-in-the-united-states/#current-eating-patterns-in-the-united-states

Dancy, J. *Ethics without Principles*. New York: Oxford University Press, 2004.

Daniel, C., Cross, A., Koebnick, C. and Sinha, R. "Trends in meat consumption in the USA." *Public Health Nutrition*, 14(4), 2011: 575–583. doi:doi:10.1017/S1368980010002077.

Darmon, N., Darmon, M., Maillot, M. and Drewnowski, A. "A nutrient density standard for vegetables and fruits: Nutrients per calorie and nutrients per unit cost." *J Am Diet Assoc.*, 105, 2005:1881–1887.

Datar, I. and Betti, M. "Possibilities for an in vitro meat production system," *Innovative Food Science and Emerging Technologies*, 1, 2010: 13–22. doi:doi:10.1016/j.ifset.2009.10.007

Deabler, Alexandra. "Arby's creates 'megetables' in response to fake meat trend: 'Why not meat-based plants?" *Fox News*, June 26, 2019. www.foxnews.com/food-drink/arbys-megetables-response-fake-meat-trend (accessed August 15, 2019).

de Boo, Jasmijn. "The future of food, why lab grown meat is not the solution." *Huffpost*, September 10, 2013. www.huffingtonpost.co.uk/jasmijn-de-boo/lab-grown-meat_b_3730367.html (accessed August 2, 2019).

DeCasien, A. R., Williams, S. A. and Higham, J. P. "Primate brain size is predicted by diet but not sociality." *Nat Ecol Evol*, 1, 2017: 0112.

Deckers, Jan. *Animal (De)liberation: Should the Consumption of Animal Products Be Banned*London: Ubiquity Press, 2016.

Dehmelt, Hans Georg. "What is the optimal anthropoid primate diet?" 2001. doi: doi:0112009

Dehmelt, Hans Georg. "Healthiest diet hypothesis." *Medical Hypotheses*, 64(4), 2005.

Delimaris, Ioannis. "Adverse effects associated with protein intake above the recommended dietary allowance for adults." *ISRN Nutrition*, 2013(126929), 2013. doi:doi:10.5402/2013/126929

DeLonge, M. "Ask a scientist union of concerned scientists, how does pollution from animal agriculture compare to vehicle pollution?" 2018. www.ucsusa.org/our-work/ucs-publications/animal-agriculture#.XFs_JC2ZPOQ.

Delvin, Hannah. "Rising global meat consumption 'will devastate environment'." *The Guardian*. www.theguardian.com/environment/2018/jul/19/rising-global-meat-consumption-will-devastate-environment (accessed August 20, 2019).

Deol, Jasraj K. and Bains, Kiran. "Effect of household cooking methods on nutritional and anti nutritional factors in green cowpea (vigna unguiculata) pods." *Journal of Food Science and Technology*, 47(5), 2010: 579–581. doi:doi:10.1007/s13197-010-0112-3

Do, Ron et al. "The effect of chromosome 9P21 variants on cardiovascular disease may be modified by dietary intake: Evidence from a case/control and a prospective study." *PLoS Medicine*, 8(10) 2011: e1001106. doi:doi:10.1371/journal.pmed.1001106

Dominick, Brian A. *Animal Liberation and Social Revolution: A Vegan Perspective on Anarchism or an Anarchist Perspective on Veganism.* Syracuse: Critical Mass Media, 1995.

Donovan, Josephine. "Feminism and the treatment of animals: From care to dialogue." *Signs*, 31(2), 2006: 305–329.

Doris, J. M. *Lack of Character: Personality and Moral Behavior.* New York: Cambridge University Press, 2002.

Ducharme, Jamie. "About 90% of Americans don't eat enough fruits and vegetables." *Time*, 2017. https://time.com/5029164/fruit-vegetable-diet/ (accessed August 3, 2019).

Duhaime-Ross, Arielle. "Test-tube burger: Lab-cultured meat passes taste test (sort of)." www.scientificamerican.com/article/test-tube-burger-lab-culture/ (accessed June 29, 2019).

Environmental Protection Agency. "What's the problem? Animal Waste Region 9 US EPA." https://www3.epa.gov/region9/water/ (accessed February 27, 2016).

Favid, S. M. "Red meat consumption and breast cancer risk." Harvard School of Public Health, 2014. www.hsph.harvard.edu/news/features/red-meat-consumption-and-breast-cancer-risk/ (accessed August 12, 2019).

Fedrigo, O., Pfefferle, A. D., Babbitt, C. C., Haygood, R., Wall, C. E. and Wray, G. A. "A potential role for glucose transporters in the evolution of human brain size." *Brain Behav. Evol.*, 78, 2011, 315–326. doi:doi:10.1159/000329852

Fessler, D. M., Arguello, A. P., Mekdara, J. M. and Macias, R.. "Disgust sensitivity and meat consumption: A test of an emotivist account of moral vegetarianism." *Appetite*, 41, 2003: 31–41.

Fitzgerald, M. *Diet Cults: The Surprising Fallacy at the Core of Nutrition Fads and a Guide to Healthy Eating for the Rest of US*. Pegasus Books, 2014.

Fleischman, Diana. "Lab meat: Survey results." *The Vegan Option Radio Show and Blog*, May 16, 2012. https://theveganoption.org/2012/05/16/lab-meat-survey-results/ (accessed August 21, 2019).

Fonseca-Azevedo, K. and Herculano-Houzel, S. "Metabolic constraint imposes tradeoff between body size and number of brain neurons in human evolution." *Proceedings of the National Academy of Sciences*, 109(45), 2012: 18571–18576. doi:doi:10.1073/pnas.1206390109

Fontana, L., Shew, J. L., Holloszy, J. O. and Villareal, D. T. "Low bone mass in subjects on a long-term raw vegetarian diet." *Arch Intern Med.*, 165(6), 2005: 684–689. doi:doi:10.1001/archinte.165. 6. 684

Foot, Philippa. *Natural Goodness*. Oxford University Press, 2001.

Francione, Gary L. *Rain without Thunder: Ideology of the Animal Rights Movement*. Philadelphia: Temple University Press, 1996.

Franks, B. Anarchism and the virtues. In Franks, B. & Wilson, M. (eds) *Anarchism and Moral Philosophy*. Palgrave Macmillan, London, 2010.

Frede, D. The historic decline of virtue ethics. In Russell, Daniel C. (ed.) *The Cambridge Companion to Virtue Ethics*. Cambridge University Press, 2013: 124–149.

Fulgoni, V. L., Keast, D. R., Bailey, R. L. and Dwyer, J. "Foods, fortificants, and supplements: Where do Americans get their nutrients?" *The Journal of Nutrition*, 141(10), 2011: 1847–1854. doi:doi:10.3945/jn.111.142257

Gajdusek, D. C. and Zigas, V. "Degenerative disease of the central nervous system in New Guinea. The endemic occurrence of "kuru" in the native population." *New England Journal of Medicine*, 257, 1957: 974–978.

Geach, Peter. *The Virtues*. Cambridge University Press, 1977.

Gerbens-Leenes, P. W., Mekonnen, M. M. and Hoekstra A. Y. "The water footprint of poultry, pork and beef: A comparative study in different countries and production systems," *Water Resources and Industry, Water Footprint Assessment (WFA) for Better Water Governance and Sustainable Development*, 1–2, 2013: 25–36. doi:doi:10.1016/j.wri.2013. 03. 001.

German, Alexander J. et al. "Dangerous trends in pet obesity." *The Veterinary Record*, 182(1) 2018: 25. doi:doi:10.1136/vr.k2

Global Environmental Change and Human Health. Science Plan and Implementation Strategy. 2019.

Goldhill, Olivia. "Scientists say your 'mind' isn't confined to your brain, or even your body." *Quartz*, 2016. https://qz.com/866352/scientists-say-your-mind-isnt-confined-to-your-brain-or-even-your-body/ (accessed August 26, 2019).

Goldhill, Olivia. "An Oxford philosopher's moral crisis can help us learn to question our instincts." *Quartz*, 2017. https://qz.com/1102616/an-oxford-philosophers-moral-crisis-can-help-us-learn-to-question-our-instincts/ (accessed July 15, 2019).

Gong, Q. and Zhang, L. "Virtue ethics and modern society – A response to the thesis of the modern predicament of virtue ethics." *Frontiers of Philosophy in China*, 5(2), 2010: 255–265.

Goodland, R. "Environmental sustainability in agriculture: Diet matters." *Ecological Economics*, 23, 1997: 189–200. doi:doi:10.1016/S0921-8009(97)00579-X

Goodland, R. and Anhang, J. "Comment to the editor." In Herrero et al. "Livestock and greenhouse gas emissions. The importance of getting the numbers right." *Animal Feed Science and Technology*, 172, 2012: 252–256. doi:doi:10.1016/j.anifeedsci.2011.12.028

Gordon, I. *Reproductive Technologies in Farm Animals*. CABI, 2004.

Gorodetsky, E. and Roberts, M. My Tongue is Meat, episode from *Two and a Half Men*. Los Angeles: CBS, 2006.

Gould, Stephen Jay. "The spices of life: An interview with Stephen Jay Gould." *Leader to Leader*, 15, The Unofficial Stephen Jay Gould Archive, 2000: 14–19. www.stephenjaygould.org/library/gould_spice-of-life.pdf (accessed August 25, 2019).

Grandin, Mary Temple. https://www.templegrandin.com (accessed August 25, 2019).

Grant, J. D. "Time for change: Benefits of a plant-based diet." *Canadian Family Physician Medecin de famille canadien*, 63(10), 2017: 744–746.

Greenhaff, L. P. "The nutritional biochemistry of creatine." *The Journal of Nutritional Biochemistry*, 8(11), 1997. doi:doi:10.1016/S0955-2863(97)00116-00112

Gruen, Lori. "Empathy and Vegetarian Commitments." In Donovan, J. and Adams, C. J. (eds) *The Feminist Tradition in Animal Ethics*, Columbia University Press, 2007: 333–344.

Gruen, Lori. *Entangled Empathy*. New York: Lantern Books, 2014.

Gupta, Charu and Dhan, Prakash. "Phytonutrients as therapeutic agents." *Journal of Complementary and Integrative Medicine*, 11(3) 2014: 151–169. doi:doi:10.1515/jcim-2013-0021

Hancox, D. "The unstoppable rise of veganism: How a fringe movement went mainstream." *The Guardian*, 2018. www.theguardian.com/lifeandstyle/2018/apr/01/vegans-are-coming-millennials-health-climate-change-animal-welfare

Harman, G. "The nonexistence of character traits." *Proceedings of the Aristotelian Society*, 100, 2000: 223–226.

Hawks, Charlotte. "How close are we to a hamburger grown in a lab?" *CNN*. www.cnn.com/2018/03/01/health/clean-in-vitro-meat-food/index.html (accessed August 8, 2019).

He, F. J., Nowson, C. A., Lucas, M. and MacGregor, G. A. "Increased consumption of fruit and vegetables is related to a reduced risk of coronary heart disease: Meta-analysis of cohort studies." *J. Hum. Hypertens*, 21, 2007: 717–728.

Health.gov. *Dietary Guidelines 2015–2020*. https://health.gov/dietaryguidelines/2015/guidelines/chapter-2/current-eating-patterns-in-the-united-states/ (accessed August 25, 2019).

Herculano-Houzel, Suzana. "Scaling of brain metabolism with a fixed energy budget per neuron: Implications for neuronal activity, plasticity and evolution." *PLoS ONE*, 6, 2011: e17514.

Herrero, M. et al. "Biomass use, production, feed efficiencies, and greenhouse gas emissions from global livestock systems." *Proceedings of the National Academy of Sciences*, 110(52), 2013: 20888–20893. doi:doi:10.1073/pnas.1308149110

Herrero, M. et al. "Livestock and the environment: What have we learned in the past decade?" *Annu. Rev. Environ. Resour.*, 40, 2015: 177–202.

Hever, Julieanna and Raymond J. Cronise. "Plant-based nutrition for healthcare professionals: Implementing diet as a primary modality in the prevention and treatment of chronic disease." *Journal of Geriatric Cardiology*, 14(5), 2017: 355–368. doi: doi:10.11909/j.issn.1671-5411.2017. 05. 012

Higdon, J. "Vitamin A." Linus Pauling Institute, Micronutrient Information Center, 2000. https://lpi.oregonstate.edu/mic/vitamins/vitamin-A

Ho, Dien. "Making ethical progress without ethical theories." *AMA Journal of Ethics*, 17 (4), 2015: 289–296.

Hocquette, J. F. "Is in vitro meat the solution for the future?" *Meat Science*, 120, 2016: 167–176. doi:doi:10.1016/j.meatsci.2016.04.036

Holdier, A. G. "The pig's squeak: Towards a renewed aesthetic argument for veganism." *Journal of Agricultural and Environmental Ethics*, 29(4), 2016: 631–642.

Hooker, Brad. Intuitions and moral theorizing. In Stratton-Lake, P. (ed.) *Ethical Intuitions and Moral Theorizing*. Oxford: Oxford University Press, 2002: 182–183.

Hursthouse, Rosalind. "Virtue theory and abortion." *Philosophy & Public Affairs*, 20(3), 1991: 223–246.

Hunt, William M. "Veganism and children: Physical and social well-being." *J. Agric. Environ Ethics*, 32, 2019: 269. doi:doi:10.1007/s10806-019-09773-4

Hursthouse, Rosalind. *On Virtue Ethics*. Oxford: Oxford University Press, 1999.

Hursthouse, Rosalind. Applying virtue ethics to our treatment of other animals. In Welchman, J. (ed.) *The Practice of Virtue: Classic and Contemporary Readings in Virtue Ethics*Indianapolis, IN: Hackett Publishing, 2006.

Hussar, K. M. and Harris, P. L. "Children who choose not to eat meat: A study of early moral decision-making." *Social Development*, 19, 2010: 627–641. doi:doi:10.1111/j.1467-9507.2009.00547.x

Ito, H., Ueno, H. and Kikuzaki, H. "Free amino acid compositions for fruits." *J. Nutr. Diet. Pract.*, 1, 2017: 1–5.

Jabbar, A., Yawar, A., Waseem, S. et al. "Vitamin B12 deficiency common in primary hypothyroidism." [Published correction appears in *J. Pak. Med. Assoc.*, 59(2), 2009: 126]. *J. Pak. Med. Assoc.*, 58(5), 2008: 258–261.

Jackson, L.S. and Al-Taher, F. Effects of consumer food preparation on acrylamide formation. In Friedman, M. and Mottram, D. (eds) *Chemistry and Safety of Acrylamide in Food. Adv. Exp. Med. Biol.*, 561, 2005. Springer, Boston, MA.

Jha, Alok. "Synthetic meat: How the world's costliest burger made it on to the plate." *The Guardian*, 2013. www.theguardian.com/science/2013/aug/05/synthetic-meat-burger-stem-cells (accessed August 20, 2019).

Jochems, Carlo E. A. et al. "The use of fetal bovine serum: Ethical or scientific problem?" *Atla-Nottingham*, 30(2), 2002: 219–228.

Johns Hopkins Bloomberg School of Public Health. "Health & environmental implications of U.S. meat consumption & production." www.jhsph.edu/research/centers-and-institutes/johns-hopkins-center-for-alivablefuture/projects/meatless_monday/resources/meat_consumptio n.html (n.d.) (accessed 3 June 2018).

Joy, Melanie. *Why We Love Dogs, Eat Pigs, and Wear Cows: An Introduction to Carnism*. Conari Press, 2001.

Joy, Melanie. "From carnivore to carnist: Liberating the language of meat." *Satya*, 18(2), 2001: 126–127.

Kant, Immanuel. *Lectures on Ethics*. New York: Harper and Row, 1963.

Kant, Immanuel. *Critique of Pure Reason*. New York: St. Martin's Press, 1929.

Kant, Immanuel. *Metaphysics of Morals*. Cambridge: Cambridge University Press, 1991.

Kass, Leon. "The wisdom of repugnance." *The New Republic*, 216(22), 1997: 17.

Kawall, J. "In defense of the primacy of the virtues." *Journal of Ethics and Social Philosophy*, 3(2), 2009: 1–21.

Kerley, Conor P. "A review of plant-based diets to prevent and treat heart failure." *Cardiac Failure Review*, 4(1), 2018: 54–61. doi:doi:10.15420/cfr.2018:1:1

Koebnick, C., Strassner, C., Hoffmann, I. and Leitzmann, C. "Consequences of a long-term raw food diet on body weight and menstruation: Results of a questionnaire survey." *Annals of Nutrition and Metabolism*, 43, 1999: 69–79. https://doi.org/10.1159/000012770

Koebnick, C., Garcia, A. L., Dagnelie, P. C., Strassner, C., Lindemans, J., Katz, N., Leitzmann, C. and Hoffmann, I. "Long-term consumption of a raw food diet is associated with favorable serum LDL cholesterol and triglycerides but also with elevated plasma homocysteine and low serum HDL cholesterol in humans." *J. Nutr.*, 135, 2005: 2372–2378.

Koehn, D. "A role for virtue ethics in the analysis of business practice." *Business Ethics Quarterly*, 5, 1995: 533–539.

Koneswaran, G. and Nierenberg, D. "Global farm animal production and global warming: Impacting and mitigating climate change." *Environmental Health Perspectives*, 116(5), 2008: 578–582. doi:doi:10.1289/ehp.11034

Kouchakoff, Paul. "The influence of food on the blood formula of man." 1st International Congress of Microbiology II. Paris: Masson & Cie, 1930: 490–493.

Kuehn, G. Dining on Fido: Death identity, and the aesthetic dilemma of eating animals. In McKenna, E. and Light, A. (eds) *Animal Pragmatism: Rethinking Human-Nonhuman Relationships*. Bloomington: Indiana University Press, 2004: 228–247.

Kunst, J. R. and Hohle, S. M. "Meat eaters by dissociation: How we present, prepare and talk about meat increases willingness to eat meat by reducing empathy and disgust." *Appetite*, 105, 2016: 758–774.

Ladouceur, Robert, Shaffer, Paige, Blaszczynski, Alex and Shaffer, Howard J. "Responsible gambling: a synthesis of the empirical evidence." *Addiction Research & Theory*, 25(3), 2017: 225–235.

Landers, T. F., Cohen, B., Wittum, T. E. and Larson, E. L. "A review of antibiotic use in food animals: perspective, policy, and potential." *Public Health Rep.*, 127(1), 2012: 4–22. doi:10.1177/003335491212700103.

Lawlor, R. "Moral theories in teaching applied ethics." *Journal of Medical Ethics*, 33(6), 2007: 370–372. doi:doi:10.1136/jme.2006.018044.

Lawton, G. "Every human culture includes cooking – This is how it began." *New Scientist*, 2016. www.newscientist.com/article/mg23230980-600-what-was-the-first-cooked-meal/ (accessed June 5, 2019).

Lebwohl, M. A. "Call to action: Psychological harm in slaughterhouse workers." *The Yale Global Health Review*, 2016. https://yaleglobalhealthreview.com/2016/01/25/a-call-to-action-psychological-harm-in-slaughterhouse-workers/ (accessed July 28, 2019).

Leischik, Roman and Spelsberg, Norman. "Vegan Triple-Ironman (raw vegetables/fruits)." *Case Reports in Cardiology*, 2014, 2014: 317246. doi:doi:10.1155/2014/317246

Link, Lilli B. et al. "Change in quality of life and immune markers after a stay at a raw vegan institute: A pilot study." *Complementary Therapies in Medicine*, 16(3) 2008: 124–130. doi: doi:10.1016/j.ctim.2008. 02. 004

Link, L. B. and Potter, J. D. "Raw versus cooked vegetables and cancer risk." *Cancer Epidemiol. Biomarkers Prev.*, 13, 2004: 1422–1435.

Liu, R. H.Health benefits of fruit and vegetables are from additive and synergistic combinations of phytochemicals." *Am. J. Clin. Nutr.*, 78, 2003: 517S-520S.

"Livestock and climate change." Worldwatch Institute. www.worldwatch.org/node/6294 (accessed April 14, 2017).

Louden, Robert. "On some vices of virtue ethics." *American Philosophical Quarterly*, 21 (3), 1984: 227–236.

Louisiana Universities Marine Consortium. "What causes ocean 'dead zones'?" *Scientific American*, 2016.

Lourenco, R. and Camilo, M. E. "Taurine: A conditionally essential amino acid in humans? An overview in health and disease." *Nutricion hospitalaria*, 17(6), 2002: 262–270.

Luke, Brian. "Justice, caring, and animal liberation." *Between the Species*, 8(2), 1992, Article 13.

Lynch, Heidi et al. "Plant-based diets: considerations for environmental impact, protein quality, and exercise performance." *Nutrients*, 10(12), 2018: 1841. doi:doi:10.3390/nu10121841

Malatesta, Errico. "Towards anarchism." Marxists Internet Archive. www.marxists.org/archive/malatesta/1930s/xx/toanarchy.htm (accessed August 26, 2019).

"Male chicks ground up alive at egg hatcheries." *CBS News*. www.cbc.ca/news/male-chicks-ground-up-alive-at-egg-hatcheries-1.823644 (accessed June 28, 2019).

Marcus, Jaqueline B. Vitamin and mineral basics: The ABCs of healthy foods and beverages, including phytonutrients and functional foods: Healthy vitamin and mineral choices, roles and applications in nutrition, food science and the culinary arts. *Culinary Nutrition*, Academic Press, 2013: 279–331. doi:doi:10.1016/B978-0-12-391882-6.00007-8

Margulis, Sérgio. "Causes of deforestation of the Brazilian Amazon." World Bank Working Paper, no. 22. Washington, DC: World Bank, 2004.

Mattick, Carolyn and Allenby, Brad. "The future of meat: Issues in science and technology." *Issues in Science and Technology*, 30(1), 2013.

Mattick, Carolyn, Landis, Amy, and Allenby, Brad. "The problem with making meat in a factory." *Slate*, 2015. https://slate.com/technology/2015/09/in-vitro-meat-probably-wont-save-the-planet-yet.html (accessed August 25, 2019).

Mattick, Carolyn S. et al., "Anticipatory life cycle analysis of in vitro biomass cultivation for cultured meat production in the United States." *Environmental Science & Technology*, 49(9), 2015: 11941–11949. doi:doi:10.1021/acs.est.5b01614

Mayo Clinic. *Patient Care and Health Info. Nutrition and Healthy Eating*, 2018. www.mayoclinic.org/healthy-lifestyle/nutrition-and-healthy-eating/in-depth/vegetarian-diet/art-20046446 (accessed July 25, 2019).

McBride, J. "B12 deficiency may be more widespread than thought." United States Department of Agriculture Agric. Res. Service, 2000. www.ars.usda.gov/news-events/news/research-news/2000/b12-deficiency-may-be-more-widespread-than-thought/ (accessed August 2, 2019).

McDowell, John. "Virtue and reason." *The Monist*, 62, 1979: 331–350.

McDowell, John. Two sorts of naturalism. In Altham, J. and Harrison, R. (eds), *Virtues and Reasons: Phillipa Foot and Moral Theory*, 1995: 149–179. New York: Oxford University Press.

Mekonnen, Mesfin M. and Hoekstra, Arjen Y. "A global assessment of the water footprint of farm animal products." *Ecosystems*, 15(3), 2012: 401–415. doi:doi:10.1007/s10021-011-9517-8

Melina, V., Craig, W., Levin, S. et al. "Position of the academy of nutrition and dietetics: Vegetarian diets." *Journal of the Academy of Nutrition and Dietetics*, 116(12), 2016: 1970–1980.

Meyer-Renschhausen, E. and Wirz, A. "Dietetics, health reform and social order: vegetarianism as a moral physiology. The example of Maximilian Bircher-Benner

(1867–1939)." *Medical History*, 43(3), 1999: 323–341. doi:doi:10.1017/s0025727300065388

Midgley, Mary. "Biotechnology and monstrosity: Why we should pay attention to the 'yuk factor'," *Hastings Center Report*, 30(5), 2000.

Milton, Katherine. "Nutritional characteristics of wild primate foods: Do the natural diets of our closest living relatives have lessons for us?" *Nutrition*, 15(6), 1999: 488–498.

Milton, Katherine. "Back to basics: Why foods of wild primates have relevance for modern human health." *Nutrition*, 16, 2000a: 481–483.

Milton, Katherine. "Hunter-gatherer diets: A different perspective." *American Journal of Clinical Nutrition*, 71, 2000b: 665–667.

"Modern vs traditional life." *Inuit Cultural Online Resource*, n.d. www.icor.inuuqatigiit.ca/explore-our-culture (accessed August 2, 2019).

Mouat, M. J. and Prince, R. "Cultured meat and cowless milk: On making markets for animal-free food." *J. Cultural Econ.*, 11, 2018: 315–329. doi:doi:10.1080/17530350.2018.1452277

Murray, Michael. *Nature Red in Tooth and Claw: Theism and the Problem of Animal Suffering*. Oxford University Press, 2008.

Najjar, Rami, Moore, Carolyn E. and Montgomery, Baxter D.. "A defined, plant-based diet utilized in an outpatient cardiovascular clinic effectively treats hypercholesterolemia and hypertension and reduces medications." *Clin. Cardiol.*, 41, 2018: 307–313.

National Institute of Health. "Iodine." 2016. https://ods.od.nih.gov/factsheets/Iodine-Consumer/#h5 (accessed August 2, 2019).

"National obesity rates & trends." NHANES. www.stateofobesity.org/obesity-rates-trends-overview/ (accessed August 22, 2019).

National Research Council (US). *Subcommittee on the Tenth Edition of the Recommended Dietary Allowances*. Recommended Dietary Allowances, 10th edn. Washington, DC: National Academies Press, 1989. www.ncbi.nlm.nih.gov/books/NBK234922/ (accessed August 2, 2019).

"New Harvest – FAQ." New Harvest. http://whyculturedmeat.org/faq/ (accessed March 26, 2019).

Notaro, K. "The crusade for a cultured alternative to animal meat: An interview with Nicholas Genovese, PhD PETA." 2011. https://ieet.org/index.php/IEET2/more/notaro20111005 (accessed October 23, 2018).

Nussbaum, Martha. "Danger to human dignity: The revival of disgust and shame in the law." *The Chronicle of Higher Education*, 50(B6), 2004.

Office of Disease Prevention and Health Promotion (ODPHP)*Dietary Guidelines 2015–2020*, n.d. https://health.gov/dietaryguidelines/2015/guidelines/chapter-2/current-eating-patterns-in-the-united-states/ (accessed December 28, 2019).

O'Malley, Keelia et al. "Vegan vs Paleo: Carbon footprints and diet quality of 5 popular eating patterns as reported by US consumers." *Current Developments in Nutrition*, 3 (1nzz047), 2019. doi:doi:10.1093/cdn/nzz047.P03-007-19

"Omega-3 fatty acids: Does your diet deliver? Most Americans don't get the recommended amount of these potentially heart-protecting fats." Harvard Medical School, 2016. www.health.harvard.edu/heart-health/omega-3-fatty-acids-does-your-diet-deliver (accessed August 22, 2019).

Oppenlander, Richard. "Freshwater abuse and loss: Where is it all going?" 2013. www.forksoverknives.com/freshwater-abuse-and-loss-where-is-it-all-going/#gs.xz97at (accessed August 25, 2019).

Oppenlander, Richard. *Food Choice and Sustainability: Why Buying Local, Eating Less Meat, and Taking Baby Steps Won't Work*. Minneapolis, MN: Langdon Street Press, 2013.

Orlich, M. J., Singh, P. N., Sabaté, J., Jaceldo-Siegl, K., Fan, J., Knutsen, S. and Fraser, G. E. "Vegetarian dietary patterns and mortality in Adventist Health Study 2." *JAMA Internal Medicine*, 173(13), 2013: 1230–1238. doi:doi:10.1001/jamainternmed.2013.6473

Oxfam. "Extreme carbon inequality." https://www-cdn.oxfam.org/s3fs-public/file_atta chments/mb-extreme-carbon-inequality-021215-en.pdf (accessed August 13, 2019).

Pachirat, Timothy. *Every Twelve Seconds: Industrialized Slaughter and the Politics of Sight*. Yale University Press, 2011.

Palmer, Clare. *Animal Ethics in Context*. Columbia University Press, 2010.

Parker, John. "The year of the vegan: Where millennials lead, businesses and governments will follow." *The Economist*, 2019 (accessed September 3, 2019).

Pem, Dhandevi and Rajesh Jeewon. "Fruit and vegetable intake: Benefits and progress of nutrition education interventions-narrative review article." *Iranian Journal of Public Health*, 44(10), 2015: 1309–1321.

PETA. "Is it OK to eat eggs from chickens I've raised in my backyard?" n.d. www.peta.org/about-peta/faq/is-it-ok-to-eat-eggs-from-chickens-ive-raised-in-my-backyard/ (accessed August 29, 2019).

Physicians Committee for Responsible Medicine (n.d.). "Health concerns about dairy: Avoid the dangers of dairy with a plant-based diet." www.pcrm.org/good-nutrition/nutrition-information/health-concerns-about-dairy (accessed August 3, 2019).

Pimentel, D. and Pimentel, M. "Sustainability of meat-based and plant-based diets and the environment." *The American Journal of Clinical Nutrition*, 78(3), 2003: 660S-663S. doi:doi:10.1093/ajcn/78.3.660S

Pincoffs, E. L. Two cheers for Meno: The definition of the virtues. In Shelp E. E. (ed.) *Virtue and Medicine*. Philosophy and Medicine, vol. 17. Dordrecht: Springer, 1985.

Plato. *Five Dialogues: Euthyphro, Apology, Crito, Meno, Phaedo*, 2nd edn. Hackett Classics, 2002.

Plumer, Brad. "Study: Going vegetarian can cut your food carbon footprint in half." *Vox*, 2016. www.vox.com/2014/7/2/5865109/study-going-vegetarian-could-cut-your-food-carbon-footprint-in-half (accessed July 10, 2019).

Rachels, James. *Created from Animals: The Moral Implications of Darwinism*. Oxford University Press, 1990.

Raha, Rosamund. Animal liberation: An interview with Professor Peter Singer." *The Vegan*, 2006: 19.

Raynor, H. A. and Epstein, L. H. "Dietary variety, energy regulation, and obesity." *Psychological Bulletin*, 127(3), 2001: 325–341. http://dx.doi.org/10.1037/0033-2909. 127.3.325 (accessed August 3, 2019).

Reclus, Élisée. "On vegetarianism." TheAnarchistLibrary.org, 2009. https://theanarchis tlibrary.org/library/elisee-reclus-on-vegetarianism (accessed August 3, 2019).

Regan, Tom. *The Case for Animal Rights*. Berkeley, CA: University of California Press, 1983.

Rehman, I. and Botelho S. Biochemistry, secondary protein structure. In *StatPearls*. Treasure Island, FL: StatPearls Publishing, 2019. www.ncbi.nlm.nih.gov/books/NBK470235/ (accessed July 13, 2019).

Ritchie, Hannah. "Which countries eat the most meat?" *BBC News*, 2019. www.bbc.com/news/health-47057341 (accessed July 13, 2019).

Rizzolatti, Giacomo and Craighero, Laila. "The mirror-neuron system." *Annual Review of Neuroscience*, 27(1), 2004: 169–192.

Rooke, J. "Do carnivores need Vitamin B12 supplements?" *Baltimore Post-Examiner*, 2013. https://baltimorepostexaminer.com/carnivores-need-vitamin-b12-supplements/2013/10/30 (accessed August 12, 2019).

Ross, L. and Nisbett, R. E. *The Person and the Situation: Perspectives of Social Psychology*. Philadelphia: Temple University Press, 1991.

Rowlands, Mark. *Can Animals Be Persons?*Oxford University Press, 2019.

Rühli, Frank, van Schaik, Katherine and Henneberg, Maciej, "Evolutionary medicine: The ongoing evolution of human physiology and metabolism." *Physiology*, 31(6), 2016: 392–397. doi:doi:10.1152/physiol.00013.2016

Rumm-Kreuter, D. and Demmel, I. "Comparison of vitamin losses in vegetables due to various cooking methods." *J. Nutr. Sci. Vitaminol.*, 36, 1990: S7–S15.

Russell, Daniel C., *The Cambridge Companion to Virtue Ethics* (Cambridge Companions to Philosophy). Cambridge University Press, 2013.

Russell, Daniel C., Virtue ethics in modern moral philosophy. In Russell, Daniel C., *The Cambridge Companion to Virtue Ethics*, Cambridge University Press, 2013: 1–7.

Ryder, Richard. *Speciesism: The Ethics of Vivisection*. Scottish Society for the Prevention of Vivisection, 1974.

Ryder, Richard. *Animal Revolution: Changing Attitudes towards Speciesism*Cambridge: Cambridge University Press, 1989.

Ryder, Richard. *Painism. A Modern Morality*. Open Gate Press, 2003.

Ryder, Richard. "Painism." In Bekoff, M. (ed.) *Encyclopedia of Animal Rights and Animal Welfare*. Santa Barbara, CA: Greenwood Press, 2010: 402–403.

Saukkonen, T., Virtanen, S., Karppinen, M. et al. "Significance of cow's milk protein antibodies as risk factor for childhood IDDM: Interactions with dietary cow's milk intake and HLA-DQB1 genotype." *Diabetologia*, 41, 1998: 72. doi:doi:10.1007/s001250050869

Schaefer, G. O. and Savulescu, J. (2014). "The ethics of producing in vitro meat." *Journal of Applied Philosophy*, 31(2), 2014: 188–202. doi:doi:10.1111/japp.12056

Schlörmann, W., Birringer, M., Bohm, V., Lober, K., Jahreis, G., Lorkowski, S., Muller, A. K., Schone, F. and Glei, M. "Influence of roasting conditions on health-related compounds in different nuts." *Food Chem.*, 180, 2015: 77–85.

"Scientists offered $1 million to grow laboratory chicken." *CNN*, n.d. www.cnn.com/2008/WORLD/americas/04/23/peta.chicken/index.html (accessed 2 Feb 2019).

Shafer-Landau, Russ. *The Ethical Life: Fundamental Readings in Ethics and Contemporary Moral Problems*, 4th edn. Oxford University Press, 2018.

Shi, Zumin et al. "Food habits, lifestyle factors and mortality among oldest old Chinese: The Chinese Longitudinal Healthy Longevity Survey (CLHLS)." *Nutrients*, 7(9), 2015: 7562–7579. doi:doi:10.3390/nu7095353

Singer, Peter. *Animal Liberation: A New Ethics for our Treatment of Animals*. New York: Avon Books, 1975.

Singer, Peter. "Utilitarianism and vegetarianism." *Philosophy & Public Affairs*, 9(4), 1980: 325–337.

Smith, Pete, Bustamante, Mercedes et al. "Agriculture, forestry and other land use (AFOLU)." www.ipcc.ch/site/assets/uploads/2018/02/ipcc_wg3_ar5_chapter11.pdf (accessed August 15, 2019).

Smithsonian National Museum of Natural History. "Introduction to human evolution." 2019. http://humanorigins.si.edu/education/introduction-human-evolution (accessed August 15, 2019).

Solomon, R. *Ethics and Excellence: Cooperation and Integrity in Business*. New York: Oxford University Press, 1992.

Specter, Michael. "The dangerous philosopher." *The New Yorker*, 1999. www.newyorker.com/magazine/1999/09/06/the-dangerous-philosopher (accessed August 12, 2019).

Specter, Michael. "Test-tube burgers." *The New Yorker*, 2019. www.newyorker.com/magazine/2011/05/23/test-tube-burgers (accessed August 12, 2019).

Steinfeld, H. et al. "Livestock's long shadow." World Wildlife Fund. www.europarl.europa.eu/climatechange/doc/FAO%20report%20executive%20summary.pdf (accessed June 14 2019).

Steinfeld, H., Gerber, P., Wassenaar, T., Castel, V., Rosales, M. and de Haan, C. "Livestock's long shadow: Environmental issues and options." Rome: FAO, 2006. www.faostat.fao.org (accessed January 19, 2017).

Stripp, C., Overvad, K., Christensen, J., Thomsen L. B., Olsen, A., Møller, S. and Tjønneland, A. "Fish intake is positively associated with breast cancer incidence rate." *The Journal of Nutrition*, 133(1), 2003: 3664–3669. doi:doi:10.1093/jn/133.11.3664

Swanton, C. The definition of virtue ethics. In Russell, Daniel C. (ed.) *The Cambridge Companion to Virtue Ethics*. Cambridge: Cambridge University Press, 2013: 315–338.

Swinburn, B., Sacks, G. and Ravussin, E. "Increased food energy supply is more than sufficient to explain the US epidemic of obesity." *The American Journal of Clinical Nutrition*, 90(6), 2009: 1453–1456. doi:doi:10.3945/ajcn.2009.28595

Tamanna, Nahid and Mahmood, Niaz. "Food processing and Maillard Reaction products: Effect on human health and nutrition." *International J. Food. Sci.*, 2015, Article ID 526762. doi:doi:10.1155/2015/526762

"Tasty foods you can feed squirrels and what to avoid." *Feeding Nature*, 2014. https://feedingnature.com/tasty-foods-you-can-feed-squirrels-and-what-to-avoid/ (accessed August 25, 2019).

"The Einstein God." *Reddit*. www.reddit.com/user/the_einsteinian_god (accessed August 28, 2019).

The Vegan Society. "Definition of veganism." www.vegansociety.com/go-vegan/definition-veganism (accessed August 23, 2019).

Thornton, Philip, Herrero, Mario and Ericksen, Polly. "Livestock and climate change." International Livestock Research Institute, 2011. https://cgspace.cgiar.org/bitstream/handle/10568/10601/IssueBrief3.pdf (accessed August 13, 2019).

"Top 5 worst celeb diets to avoid in 2018." The British Dietetic Association (BDA), 2017. www.bda.uk.com/news/view?id=195 (accessed August 27, 2019).

Tuomisto, L. H. "Importance of considering environmental sustainability in dietary guidelines." *The Lancet Planetary Health*, 2(8), 2018. doi:doi:10.1016/S2542-5196(18)30174-30178

Tuomisto, L. H. and Joost Teixeira de Mattos, M. "Environmental impacts of cultured meat production." *Environmental Science & Technology*, 45(14), 2011: 6117–6123. doi:doi:10.1021/es200130u

Turner-McGrievy, G., Mandes, T. and Crimarco, A. "A plant-based diet for overweight and obesity prevention and treatment." *Journal of Geriatric Cardiology*, 14(5), 2017: 369–374. doi:10.11909/j.issn.1671-5411.2017. 05. 002

Tuso, P. J., Ismail, M. H., Ha, B. P. and Bartolotto, C. "Nutritional update for physicians: Plant-based diets." *The Permanente Journal*, 17(2), 2013: 61–66. doi:doi:10.7812/TPP/12-085

Ulvik, A., Vollset, S. E., Hoff, G. and Ueland, P. M. "Coffee consumption and circulating b-vitamins in healthy middle-aged men and women." *Clin. Chem.*, 54, 2008: 1489–1496.

University of Southern California. "Meat and cheese may be as bad for you as smoking." *ScienceDaily*. www.sciencedaily.com/releases/2014/03/140304125639.htm (accessed June 20, 2018).

UNPD (United Nations Population Division). *The 2006 Revision and World Urbanization Prospects: the 2005 Revision*. Population Division of the Department of Economic and Social Affairs of the United Nations Secretariat, World Population Prospects. See http://esa.un.org/unpp (accessed June 20, 2018).

Vanderpump, M. P. and Turnbridge, W. M. "Epidemiology and prevention of clinical and subclinical hypothyroidism." *Thyroid*, 12, 2002: 839–847.

Van Der Zee, Bibi. "What is the true cost of eating meat?" *The Guardian*, 2018. www.theguardian.com/news/2018/may/07/true-cost-of-eating-meat-environment-health-animal-welfare (accessed August 22, 2019).

Vegetables and fruits. In Harvard Medical School, *The Nutrition Source*, 2019. www.hsph.harvard.edu/nutritionsource/what-should-you-eat/vegetables-and-fruits/ (accessed August 24, 2019).

Verbeke, Wim et al. "Would you eat cultured meat? Consumers reactions and attitude formation in Belgium, Portugal and the United Kingdom." *Meat Sci.*, 102, 2015: 49–58. doi:10.1016/j.meatsci.2014. 11. 013

Vieira, Samantha A. et al. "Challenges of utilizing healthy fats in foods." *Advances in Nutrition*, 6(3) 2015: 309S-317S. doi:doi:10.3945/an.114.006965

Vining, Alexander Q. and Nunn, Charles L. "Evolutionary change in physiological phenotypes along the human lineage." *Evolution, Medicine, and Public Health*, 2016(1), 2016: 312–324. doi:doi:10.1093/emph/eow026

"Vitamin B12: What to know." *WebMD*, 2019. www.webmd.com/diet/vitamin-b12-deficiency-symptoms-causes#1 (accessed August 15, 2019).

"Vitamin D." National Institutes of Health, 2019. https://ods.od.nih.gov/factsheets/VitaminD-HealthProfessional/ (accessed August 26, 2019).

Vogel, Gretchen. "Organs made to order." *Smithsonian Magazine*, 2010. www.smithsonianmag.com/science-nature/organs-made-to-order-863675/ (accessed August 3, 2019).

Wacker, M. and Holick, F. M. "Sunlight and Vitamin D." *Dermato-Endocrinology*, 5(1), 2013: 51–108. doi:doi:10.4161/derm.24494

Walsh, B. "The triple whopper environmental impact of global meat production." *Time*, 2013. http://science.time.com/2013/12/16/the-triple-whopper-environmental-impact-of-global-meat-production/ (accessed June 13, 2018).

Walsh, Stephen P. K. Why foods derived from animals are not necessary for human health. In Linzey, C. (ed.) *Ethical Vegetarianism and Veganism*. Oxon: Routledge, 2018: 19–33. doi:doi:10.4324/9780429490743-2

Wang, Y.C., McPherson, K., Marsh, T., Gortmaker, S.L. and Brown, M. "Health and economic burden of the projected obesity trends in the USA and the UK." *Lancet*, 378(9793), 2011: 815–825.

Warren, Karen. "The promise and power of ecofeminism." *Environmental Ethics*, 12(2), 1990: 125–146.

Warren, Mary Anne. "Difficulties with the strong rights position." *Between the Species*, 2 (4), 1987: 433–441.

Watts, S. "First they tortured animals, then they turned to humans." *A&E*. www.aetv.com/real-crime/first-they-tortured-animals-then-they-turned-to-humans,2018 (accessed June 18, 2019).

Webb, Marryn Somerset. "The veganism boom does more for food company profits than the planet ." *Financial Times*, 2019. www.ft.com/content/79c2aa16-35f1-11e9-bb0c-42459962a812 (accessed August 27, 2019).

"What are proteins and what do they do?" U.S. National Library of Medicine, 2019. https://ghr.nlm.nih.gov/primer/howgeneswork/protein (accessed August 29, 2019).

"When do kids realize that eating meat involves killing animals?" *Quora*. www.quora.com/When-do-kids-realize-that-eating-meat-involves-killing-animals (accessed June 27, 2018).

"Why cultured meat." http://whyculturedmeat.org/essays/animal-rights/is-it-animal-rights/ (accessed August 12, 2019).

Wicker, Bruno, Keysers, Christian, Plailly, Jane, Royet, Jean-Pierre, Gallese, Vittorio and Rizzolatti, Giacomo. "Both of us disgusted in my insula: The common neural basis of seeing and feeling disgust." *Neuron*, 40(3), 2003: 655–664.

Wicks D. "Silence and denial in everyday life: The case of animal suffering." *Animals: An Open Access Journal from MDPI*, 1(1), 2011: 186–199. doi:doi:10.3390/ani1010186

Wittgenstein, Ludvig. *Philosophical Investigations*. Oxford: Blackwell, 1953.

Women Champion Peace & Justice through Nonviolence. (n.d.) www.library.georgetown.edu/exhibition/women-champion-peace-justice-through-nonviolence (accessed June 12, 2019).

Wood, Allen W. *Kantian Ethics*. Cambridge University Press, 2018.

Woolf, C. J. and Ma, Q. "Nociceptors: Noxious stimulus detectors." *Neuron*, 55, 2007: 353–364.

World Health Organization. "Availability and changes in consumption of animal products." 2015. www.who.int/nutrition/topics/3_foodconsumption/en/index4.html (accessed August 12, 2019).

"World's first lab-grown burger is eaten in London." *BBC News*, 2013. www.bbc.com/news/science-environment-23576143 (accessed August 24, 2019).

Zemeckis, Robert (dir.). *Contact*. Burbank, CA: Warner Home Video, 1997. DVD.

Zhong, V.W., Van Horn, L., Cornelis, M. C., Wilkins, J. T., Ning, H., Carnethon, M. R., Greenland, P., Mentz, R. J., Tucker, K. L., Zhao, L., Norwood, A. F., Lloyd-Jones, D. M. and Allen, N. B. "Associations of dietary cholesterol or egg consumption with incident cardiovascular disease and mortality." *JAMA*, 19(11), 2019: 1081–1095. doi:doi:10.1001/jama.2019.1572

Zink, Katherine D. and Lieberman, Daniel E.Impact of meat and Lower Palaeolithic food processing techniques on chewing in humans." *Nature*, 531, 2016: 500–503.

Index

Abortion 55
Acrylamides 4, 78, 81, 127
Alienation 52, 103
American Cancer Society 128
American Dietetic Association 64, 108
Anarchism 103, 104
Aquinas, Thomas 36, 37
Arby's 85
Aristotle 18, 19, 20, 21

B12 5, 64, 65, 110, 111
Bareburger 84
Beyond Meat 84
Bircher-Benner, Maximilian 69
British Dietetic Association 80

Cancer 63, 64, 78, 113, 114, 121, 128
Cancer Council 113
Cannibalism 45, 54, 55, 56
Cartesian 13
Centers for Disease Control and Prevention 1
Children i, vii 5, 19, 31, 35, 53, 94, 95, 99, 107–113, 115, 116, 117, 119–123
Christianity 18, 119
Chromosome 9p21, 78
Compassion 8, 18, 45, 52, 56, 93, 94, 95, 96, 102, 104
Compassionate 24, 93, 95, 96
Cooking ix, 2–5, 38–40, 65, 69, 70–73, 77, 78, 79, 81, 83, 100, 111, 126, 127, 128
Coronary Heart Disease (CHD) 74
Cruel 34, 37, 98
Cruelty 19, 33, 36, 37, 46, 48, 55, 60, 93, 95, 99

Deckers, Jan 13, 35, 90
Degradation 3, 25, 33, 34, 37, 43, 51, 52, 57, 65, 66, 73, 89–91, 94, 96, 100, 102, 121, 125
Denaturation, see Denatures
Denatures ix, 4, 78, 79, 127
Deontic, see Deontology
Deontological, see Deontology
Deontologist, see Deontology
Deontology 3, 17, 22, 23, 24, 28, 45, 46, 55–57
Descartes, René 3, 32
Diabetes 64, 69, 82, 114, 121
Diet i, iii, vii, ix 1–5, 8, 33, 34, 44, 46, 50, 52, 53, 57, 60–67, 69, 70–84, 89, 91, 92, 93, 94, 99, 100, 101–104, 107–117, 119–122, 123, 126–129
Diets Dietary, Dietetic, Dietetics, See Diet
Disease ix, 1, 2, 31, 33, 35, 52, 53, 54, 62, 64, 66, 69, 73, 74, 78, 90, 91, 93, 108, 109, 111, 113, 114, 121, 122, 128
Donovan, Josephine 31, 97

Education i, vii 5, 84, 89, 94–97, 99, 101, 121
Empathy 94, 122, 123
Encephalization 70, 72, 73
Environment i, ii, viii, ix, x 3–5, 8, 10, 11, 14, 22, 25, 28, 32–34, 37, 43–47, 50–52, 55, 57, 60, 65, 66, 69, 76, 83–85, 89–94, 96, 98, 100–103, 115, 121, 123, 126–128
Environmental, See Environment
Environmentalist, See Environment
Environmentally, See Environment
Epidemic 91
Evolution viii, 3, 33, 70, 71, 127

Index

Fat 2, 47, 50, 61, 63, 66, 80, 114
Flavor 37, 38, 39, 81,
Freedom 52, 69, 96, 104
Fruit, ix, 1–5, 31, 34, 38, 46, 50, 53, 54, 56, 62, 63, 65, 66, 67, 72–78, 80, 81, 82, 83, 93, 94, 100, 101, 102, 109, 110, 113, 115, 122, 127, 128

Global warming 32, 33
Gould, Stephen Jay 77
Gruen, Lori 97

Happiness 6, 7, 8, 19–23
Heart disease 84, 88, 114, 121
Homo erectus 69
Homo sapiens 47, 69, 70, 76
Hursthouse, Rosalind 17, 18, 55

Impossible (Burger) 84
Influenza 91
In vitro 4, 14, 43, 44, 46, 54, 55, 56

Jain 117, 119
Justice 18, 45, 46, 96, 103, 104

Kant, Immanuel 10, 22, 24, 36, 37, 46, 98
Kantian, see Kant, Immanuel
Kass, Leon 55, 56

Lab-grown 50, 57, 67
Legal ban 90, 94, 101, 102
Leukocytosis 4, 127
Livestock 52, 90–94, 96, 101, 102, 103, 111, 128

MacDonald's 84
Machupo 91
Maillard Reaction 78, 84
Megetable 85
Midgley, Mary 29, 55
Moderation 2, 3, 4, 6, 17, 18, 19, 21, 46, 54, 64, 66, 67, 74, 81

National Cancer Institute 128
National Health and Nutrition Examination Survey 1
Newkirk Ingrid 57, 62
Nonviolence 4, 6, 17, 19, 36, 66
Nussbaum, Martha 55

Obesity ix, 1, 2, 69, 114
Office of Disease Prevention and Health Promotion 109

Omega-3 65, 66, 101
Oxfam 89

Pain 4, 6, 7, 8, 11, 12, 13, 16, 22, 23, 28, 32, 36, 40, 54, 99, 100
People for the Ethical Treatment of Animals (PETA) 57, 61, 62
Phytonutrients 78
Plant-based 37, 40, 48, 49, 51, 57, 61, 62, 63, 73, 74, 77, 83, 91, 94, 101, 102, 103, 114
Population 1, 2, 53, 83, 91, 98, 109, 110, 111, 115, 128
Protein/proteins 4, 49, 53, 64, 67, 78, 81, 83, 92, 101, 109, 127
Psychology of disgust 30

Regan, Tom 6, 9, 10, 11, 28, 46
Repugnance, see Wisdom of repugnance
Rights 3, 4, 6, 7, 9, 10, 11, 15, 17, 18, 19, 20, 21, 23, 24, 28, 30, 32, 45, 46, 49, 51, 55, 62, 65, 90, 92, 101, 117, 120, 129

Search for Extraterrestrial Intelligence (SETI Institute) 129
Self-indulgence 84, 46, 52, 54, 56
Sentience 4, 12, 13, 14
Sentient 4, 7, 9, 12, 13, 22, 32, 40, 80
Singer, Peter 6, 8, 9, 13, 15, 22, 28
Sodium 81
Speciesism 9, 102
Subject-of-a-life 10, 29
Suffering vii, viii 3, 6, 7–14, 22, 28–30, 34, 36, 37, 40, 43, 44, 46–52, 55, 57, 89, 90, 99, 101, 103, 104, 121, 123
Sugar ix, 2, 4, 63, 127

Temperance 18, 21, 45, 53, 54, 56, 96
TGI Friday's 84
Tofurkey 119, 120, 51

Undemocratic 96
Utilitarian 3, 6, 7, 8, 9, 11, 15, 17, 18, 22, 23, 24, 28, 29, 45
Utilitarianism, see Utilitarian

Vegan i, viii, ix 2–8, 12, 15, 22, 28, 29, 33, 37, 40, 43, 44, 46, 47, 49, 51, 56, 57, 60, 61–67, 69, 71, 74, 75, 78–85, 89–100, 102–105, 107–113, 115–123, 127–129
Veganarchism ix, 83, 103, 104
Veganism, see Vegan

Vegan project 90
Virtue viii, 4, 6, 11, 15, 17, 18, 19, 20, 21, 22, 23, 24, 25, 28, 29, 34, 40, 45, 47, 48, 51, 55, 66, 67, 79, 83, 84, 95, 98, 104, 128
Virtue ethics 6, 15, 17, 18–22, 24, 84, 104
Vitamin 64, 65, 66, 77, 78, 79, 81, 110, 111, 112
Vitamin D 66, 78, 112

White blood cells 4, 79, 127
White Castle 84, 102
Wisdom of repugnance 30, 55, 56
Wittgenstein 13
World Health Organization 63

Yuck factor 29, 55

Zoonoses ix
Zoonotic 90, 91